Part L explained —
The BRE Guide

BRE publications relating to Part L

Achieving airtightness
This three-part Good Building Guide gives practical advice on achieving airtightness in new buildings. GG67, 2006

Airtightness in commercial and public buildings
This design guide sets out the principles of providing an effective airtightness layer and advises on common pitfalls. BR448, 2002

Assessing the effects of thermal bridging at junctions and around openings
Guidance on assessing the effects of thermal bridging at junctions and around openings in the external elements of buildings on overall heat loss or gain. IP1/06, 2006

Conventions for U-value calculations
The new edition of this guide indicates which methods of U-value calculation are appropriate for different construction types, with data for typical constructions. BR443, 2006

Selecting lighting controls
Explains how to choose lighting controls to take account of the space, its use and the daylight available. Describes the common types of control and how to calculate energy savings. DG498, 2006

Site layout planning for daylight and sunlight
This important and widely used guide provides advice on the planning of the external environment. BR209, 1998

Solar shading of buildings. Types of shading devices and special types of glazing are discussed and their comparative performance summarised. BR 364, 1999

Summertime solar performance of windows with shading devices
Provides data for quantifying the ability of windows and shading devices to control summertime overheating. Includes a calculation tool on CD-ROM. FB9, 2005

Thermal insulation: avoiding risks
BRE recommendations on good design and construction practice associated with thermal standards. BR262, 2001

Full details of all BRE publications are at www.brepress.com

Part L explained —
The BRE Guide

BRE is committed to providing impartial and authoritative information on all aspects of the built environment for clients, designers, contractors, engineers, manufacturers and owners. We make every effort to ensure the accuracy and quality of information and guidance when it is published. However, we can take no responsibility for the subsequent use of this information, nor for any errors or omissions it may contain.

BRE is the UK's leading centre of expertise on the built environment, construction, sustainability, energy, fire and many associated issues. Contact BRE for information about its services, or for technical advice:
BRE, Garston, Watford WD25 9XX
Tel: 01923 664000, enquiries@bre.co.uk
www.bre.co.uk

BRE publications are available from
www.brepress.com
or
IHS ATP (BRE Press)
Willoughby Road
Bracknell RG12 8FB
Tel: 01344 328038, Fax: 01344 328005
brepress@ihsatp.com

Requests to copy any part of this publication should be made to the publisher:
BRE Press
Garston, Watford WD25 9XX
Tel: 01923 664761, brepress@emap.com

BR 489
© Copyright BRE 2006
First published 2006
ISBN 1 86081 910 9

Contents

Preface		viii
Foreword		ix
1	**Introduction**	1
	The Building Regulations Part L	2
	Reducing carbon dioxide emissions from buildings	2
	Energy White Paper	5
	Building Regulations commitments	6
2	**Review of Part L**	7
	2004 consultation	7
	Impact assessment	7
	New buildings	8
	Existing building stock	8
	Controlled fittings and services, and thermal elements	9
	Consequential improvements	10
	Review of sustainable buildings	10
	Measures for improving compliance	11
	Competent person schemes	12
3	**Energy Performance of Buildings Directive**	15
	Requirements	15
4	**Implementing the changes to Part L**	17
	Domestic boiler amendment – April 2005	17
	Main amendment – April 2006	17
	EPBD Articles 3 to 6	17
	Low and zero carbon systems	18
	EPBD Articles 7 to 10	18
	Certification of energy performance	18
5	**National Calculation Methodology**	23
	Carbon dioxide targets	23
	Calculating carbon dioxide emissions	24
	Standard Assessment Procedure (SAP)	24
	Simplified Building Energy Model (SBEM)	26
6	**Part L Regulations and approved guidance**	27
	Statutory Instruments	27
	Definitions	27
	Regulations	31
	Exempt buildings and work	34
	Transitional arrangements	34

	Requirements in Schedule 1	35
	Guidance on complying with the Regulations	36
	Approved Documents	36
	Second tier guidance	37
	Changes to Part L introduced in 2006	37
7	**Construction of new buildings**	**41**
	General guidance	41
	Types of work covered	41
	Demonstrating compliance	41
	Design standards	42
	Criterion 1 – Predicted carbon dioxide emission rate to be no greater than target	42
	Criterion 2 – Performance to be within design limits	49
	Criterion 3 – Building to have passive control measures to limit solar gain	54
	Quality of construction and commissioning	56
	Criterion 4 – As-built performance to be consistent with design	56
	Providing information	61
	Criterion 5 – Information to be provided for energy efficient operation	61
	Model designs	62
8	**Work in existing buildings**	**63**
	General guidance	63
	Types of work covered	63
	Guidance relating to building work	66
	Work on controlled fittings	66
	Work on controlled services	66
	Guidance on thermal elements	72
	New and replacement thermal elements	72
	Renovation of thermal elements	73
	Retained thermal elements	74
	Extension of a building	74
	Fabric and building services standards	75
	Conservatories and substantially glazed extensions	76
	Large extensions to non-domestic buildings	77
	Optional approaches with more design flexibility	77
	Change of energy status and material change of use	78
	Material alteration	79
	Consequential improvements	79
	Extensions	79
	Building services	80
	Providing information	81

Annex 1: Standard Assessment Procedure (SAP) — 83
- SAP rating — 83
- Environmental Impact rating — 83
- Dwelling carbon dioxide Emission Rate (DER) — 84
- Calculation method — 84
- Scope — 85

Annex 2: Simplified Building Energy Model — 87
- Introduction — 87
- Energy calculation tools — 87
- Target carbon dioxide emissions — 88
- Energy Performance of Buildings Directive — 88
- National Calculation Methodology — 88
- SBEM — 89

Annex 3: Avoiding overheating — 91
- Dwellings — 91
- Buildings other than dwellings — 92

Annex 4: Airtightness testing — 94
- Introduction — 94
- Part L air permeability requirements — 95
- Part L airtightness testing requirements — 95
- Guidance — 95

Annex 5: Low and zero carbon systems — 96
- Absorption cooling — 98
- Biomass heating — 99
- Micro-CHP — 99
- Ground cooling — 100
- Ground source heat pumps — 100
- Solar electricity — 101
- Solar hot water — 101
- Wind turbines — 102

Annex 6: Domestic heating compliance guide — 103
- Introduction — 103

Annex 7: Non-domestic heating, cooling and ventilation compliance guide — 105
- Introduction — 105

Abbreviations — 107
References — 109

Preface

This book is written as an introductory guide to Part L of the Building Regulations that came into effect in April 2006. The changes to the Regulations and the Approved Documents are complex and wide-ranging, and it is hoped that all those who commission, specify and construct new buildings and adapt existing buildings will find it valuable in 'getting up to speed' in the new regime.

It does not replicate the guidance in the Approved Documents, but highlights the key requirements, provides further explanation where desirable, and explains the differences between the requirements for dwellings and other buildings.

The Approved Documents should be consulted for full details of any aspects that are covered only briefly in the guide, and the ODPM website *www.odpm.gov.uk/building-regulations* should be reviewed for new information updating the Regulations, and also for information about some of the 'second tier' documents that were not published at the time of going to press.

Drafting and editing of the book have been under the overall control of Ken Bromley, BRE Environment, who was seconded to ODPM Buildings Division during 2004 and 2005 to work with the team revising Part L. Preparation of the book has drawn on the knowledge of many BRE staff with particular expertise in certain areas relating to energy efficiency and the Building Regulations, in particular:

Brian Anderson (U-values, SAP)
Roger Hitchin (SBEM)
Mike Jaggs (airtightness)
Paul Littlefair (lighting).

The cooperation and support of ODPM Buildings Division in the preparation of this guide is gratefully acknowledged.

Foreword

Our existing buildings have a major impact on the environment and our new buildings leave a legacy that can last for many generations.

The 2003 Energy White Paper identified energy efficiency as the "cheapest, cleanest and safest way" of delivering the Government's energy and environmental policy objectives. Bringing forward the revision of Part L, and implementing the EU Energy Performance of Buildings Directive, were key commitments in the Government's Energy Efficiency 'Plan for Action'.

Energy efficient buildings can deliver major benefits in terms of both reduced environmental impact and lower operating cost. However, they must also be healthy, safe and productive. Integrated, intelligent design delivers reduced construction cost and healthy, productive homes and workplaces. The new Part L requirements require much closer collaboration between the building client and all members of the design team. Key design decisions associated with form, fabric, façade, HVAC system and lighting, etc are now inextricably linked.

The changes to Part L are radical and far reaching. This guide is designed to help designers and builders through the maze and to provide clear guidance in achieving cost-effective compliance with the new requirements.

With the introduction of mandatory pressure testing and competent person schemes for calculating Carbon Emission Rates, Part L compliance is likely to become more demanding. Also, tougher enforcement, coupled with building labelling/ certification, brings with it new risks and potential liabilities for designers and builders — this guide will help to reduce risk and deliver buildings with substantially lower energy requirements with major benefits in terms of reduced energy cost and environmental impact.

Part L 2006 will make architectural 'greenwash' much more difficult and building energy labelling in the future is likely to introduce an important new driver for building clients, owners and operators associated with their brand equity and corporate social responsibility positioning. Clients are increasingly likely to demand buildings that exceed the minimum energy performance requirements dictated by Part L — this guide will help you respond to these new challenges.

David Strong
Managing Director, BRE Environment
Chairman of the UK advisory group on
implementation of the EPBD

1 Introduction

This guide has been prepared to help architects and builders understand the new energy performance requirements in Part L of the Building Regulations, *Conservation of fuel and power*. To satisfactorily deliver these new requirements, much closer collaboration will be needed between building clients and all members of the design team.

The guide covers the major changes to Part L introduced in April 2006, including:
- the background to the changes
- the EU Energy Performance of Buildings Directive
- the Regulations and approved guidance that implement the changes
- designing buildings to meet the new carbon dioxide emission targets
- new standards for work in existing buildings.

In addition, annexes cover:
- the new tools for calculating carbon dioxide emissions from buildings – SAP 2005 and SBEM
- limiting the effects of overheating in buildings
- air pressure testing
- low and zero carbon energy sources
- introductions to the *Domestic heating compliance guide* and *Non-domestic heating, cooling and ventilation compliance guide*.

The Building Regulations are made under powers provided in the Building Act 1984, and apply in England and Wales. The current edition is *The Building Regulations 2000*, as amended (most recently in April 2006), and most building projects must comply with them. Separate regulations apply in Scotland and Northern Ireland.

The formal 'Requirements' of the Building Regulations are listed in a schedule (Schedule 1) to the Building Regulations 2000 and are grouped into 14 Parts. Further information about the Regulations is available in a booklet published by the Office of the Deputy Prime Minister[1].

The Building Regulations Part L

Part L of the Building Regulations sets energy efficiency rules that apply to the construction of new buildings, and to certain works associated with extending, altering and changing the use of existing buildings.

> Part L does not specify the energy-saving devices or materials to be used, but instead sets levels of required energy performance

Requirements in the Building Regulations are generally functional rather than prescriptive: so Part L does not specify the energy-saving devices or materials to be used, but instead sets levels of required energy performance. This allows designers the flexibility to innovate and to choose the most cost-effective and practical solutions.

Guidance showing some ways of meeting the functional requirements is published in so-called 'Approved Documents' – see box on next page. For Part L there are four Approved Documents:

- *Conservation of fuel and power in new dwellings* (ADL1A)
- *Conservation of fuel and power in existing dwellings* (ADL1B)
- *Conservation of fuel and power in new buildings other than dwellings* (ADL2A)
- *Conservation of fuel and power in existing buildings other than dwellings* (ADL2B).

Reducing carbon dioxide emissions from buildings

Carbon dioxide is the main greenhouse gas in the UK responsible for global warming, and around 50% of all carbon dioxide emissions are associated with energy use in buildings.

To help reduce UK energy consumption and carbon dioxide emissions, energy performance standards for buildings have been

---

[1] *Building Regulations Explanatory Booklet*, ODPM, 2006. Available from *www.odpm.gov.uk*

progressively raised over the years, with major amendments to Part L appearing in 1990, 1995, 2002 and – most recently – April 2006.

The revisions to Part L in 2002 reduced permitted carbon dioxide emissions from new buildings by around 25%. The revisions in 2006 reduced emissions further – by an average of 20% for new dwellings, 23% for naturally ventilated buildings, and up to 28% for air-conditioned buildings, bringing the total improvement since before 2002 to over 40%.

Building Regulations Approved Documents

A: Structure. 2004 edition incorporating 2004 amendments
B: Fire safety. 2000 edition incorporating 2000 and 2002 amendments
C. Site preparation and resistance to contaminants and moisture. 2004 edition
D: Toxic substances. 1992 edition incorporating 2002 amendments
E: Resistance to the passage of sound. 2003 edition incorporating 2004 amendments
F: Ventilation. 2006 edition
G: Hygiene. 1992 edition incorporating 1992 and 2000 amendments
H: Drainage and waste disposal. 2002 edition
J: Combustion appliances and fuel storage systems. 2002 edition
J: 2002 Edition: Guidance and supplementary information on the UK implementation of European standards for chimneys and flues
K: Protection from falling, collision and impact. 1998 edition incorporating 2000 amendments
L1A: Conservation of fuel and power in new dwellings. 2006 edition
L1B: Conservation of fuel and power in existing dwellings. 2006 edition
L2A: Conservation of fuel and power in new buildings other than dwellings. 2006 edition
L2B: Conservation of fuel and power in existing buildings other than dwellings. 2006 edition
M: Access to and use of buildings. 2004 edition
N: Glazing – safety in relation to impact, opening and cleaning. 1998 edition incorporating 2000 amendments
P: Electrical safety – dwellings. 2006 edition
Approved Document to support regulation 7: Materials and workmanship. 1992 edition incorporating 2000 amendments

Figure 1 shows how carbon dioxide emissions associated with gas-fired space and water heating in a typical new semi-detached house (floor area 80 m²) have fallen – from 4.4 tonnes per year in 1981 to 1.7 tonnes per year in 2006.

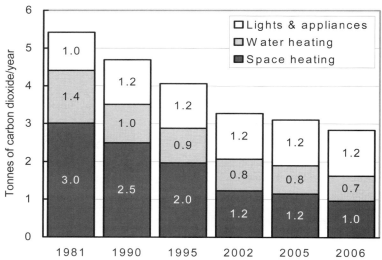

Figure 1 Carbon dioxide emissions from a typical new semi-detached house (80 m²)

Over the same period, carbon dioxide emissions associated with lights and electrical appliances in a typical house – which apart from fixed lights are not controlled by Building Regulations – have risen from 1.0 to 1.2 tonnes per year.

Figure 2 shows carbon dioxide emissions from a new air-conditioned office, for comparison purposes with the same floor area of 80 m². Emissions fell between 1995 and 2006 from 6.0 to 3.3 tonnes of carbon dioxide per year. Figures are given for gas-fired heating and domestic hot water, and electrically-powered air conditioning, lighting and auxiliary equipment.

To put these figures into perspective, a family car doing 12,000 miles a year at 33 miles to the gallon emits 3.9 tonnes of carbon dioxide. And a family of four breathes out around 1.5 tonnes of carbon dioxide per year!

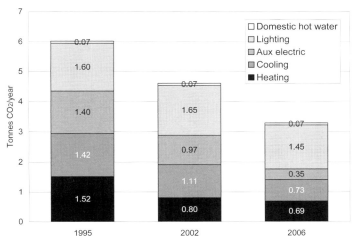

Figure 2 Carbon dioxide emissions from a new air-conditioned office (80 m^2)

Energy White Paper

Work on the amendment to Part L that came into force in April 2006 began in February 2003 when the Government set out its energy policy in an Energy White Paper Our energy future – creating a low carbon economy. It addressed the four key issues of:
- the environment
- reliability of energy supply
- affordable energy for the poorest
- competitive energy markets for businesses, industries and households.

The White Paper put the UK on a path to a 60% reduction in carbon dioxide emissions by 2050 – from around 605 to 238 million tonnes of carbon dioxide (MtCO$_2$) per year[2].

The UK already had a Kyoto Protocol commitment to reduce a basket of greenhouse gas emissions by 12.5% below 1990 levels by 2008–12, and a national goal under the Climate Change

> The Energy White Paper put the UK on a path to a 60% reduction in carbon dioxide emissions by 2050

---

[2] The Energy White Paper uses units of tonnes of carbon rather than tonnes of carbon dioxide. 1 tonne of carbon is equivalent to 3.67 tonnes of carbon dioxide.

Programme[3] to move towards a 20% reduction in carbon dioxide emissions – 73 MtCO$_2$ per year – below 1990 levels by 2010. The White Paper indicated that further cuts of 55–92 MtCO$_2$ per year would be needed by 2020 to remain on track for a 60% reduction by 2050.

Building Regulations commitments

To help in achieving its carbon reduction objectives, the Government made a number of commitments on Building Regulations in the Energy White Paper:

- raise Part L standards for new and existing buildings (including a specific requirement that new and replacement boilers in homes should be of the high efficiency, condensing type)
- improve compliance with Part L
- implement the Energy Performance of Buildings Directive
- bring forward the next amendment of Part L and aim to bring into effect in 2005.

---

[3] Following an extensive review, the Government published a new UK Climate Change Programme on 28 March 2006 – see *www.defra.gov.uk/environment/ climatechange*

2 Review of Part L

After the Energy White Paper was published in 2003, the Government department responsible for the Building Regulations, the Office of the Deputy Prime Minister (ODPM), began a review of Part L requirements to see how standards could be improved, and how the White Paper commitments could be implemented.

2004 consultation

Following extensive discussions with stakeholders, ODPM published a consultation paper in July 2004 containing wide-ranging proposals for amending Part L[4].

The paper included four draft Approved Documents – L1A, L1B, L2A and L2B – giving guidance for new dwellings, existing dwellings, new non-domestic buildings and existing non-domestic buildings, respectively.

At the end of a three-month consultation period, the Government analysed the responses and adjusted the proposals to ensure that they would be cost-effective and practical.

Impact assessment

Energy efficiency standards had already been improved by 25% in the 2002 revisions, and any further changes – as for all Building Regulations – would need to be demonstrably cost-effective, practicable, proportionate and flexible, and not introduce undue technical risks.

Proposals to amend regulations must be supported by a Regulatory Impact Assessment showing that the benefits of

---

[4] *www.odpm.gov.uk/building-regulations*

the proposed changes will outweigh the costs. For the 2006 revision of Part L, for the first time, ODPM took account of the 'social cost' of carbon (valued at around £75 per tonne), as a result of which more measures for reducing carbon emissions became cost-effective.

For measures to be cost-effective, the marginal (that is extra) capital and labour costs must be balanced by the present value of both the energy savings and the avoided social costs over the life of the measure (for example 40 years for windows and 60 years for the building fabric).

UK manufacturing industry also had to be in a position to supply materials to proposed new standards. This influenced, for example, the U-value requirements that were introduced for windows. To give manufacturers time to build up production of higher performance windows with soft, low-emissivity coatings, requirements for replacement windows in existing dwellings were left unchanged, although standards for new windows in extensions were raised.

New buildings

To implement the Energy Performance of Buildings Directive (see Chapter 3), standards for new buildings would need to be in terms of whole building targets for carbon dioxide emissions.

> This approach gives designers and builders flexibility to choose solutions that best meet their needs

This approach does give designers and builders the flexibility to choose solutions that best meet their needs. However, limits need to be set on design flexibility:

- first, to avoid technical problems such as condensation on surfaces with poor U-values, and
- second, to discourage excessive and inappropriate trade-off – for example the design of buildings with poor insulation standards, offset by renewable energy systems that might be replaced in a few years by conventional building services.

Existing building stock

The review of Part L paid particular attention to measures that could improve the energy efficiency of the existing building stock, which is poor by comparison with new buildings.

For example, the average energy efficiency of houses built before 2002 is only half that of houses built to post-2002 standards. But each year less than 0.1% of the existing stock is demolished[5], so the energy efficiency of existing as well as new buildings needs to be improved if carbon emissions from the whole of the stock are to fall.

> The energy efficiency of the existing building stock is poor by comparison with new buildings

Figure 3 illustrates how, taking account of both construction and demolition rates, the estimated proportion of UK dwellings built to 2002 standards or better in the existing stock will increase only slowly through to 2050.

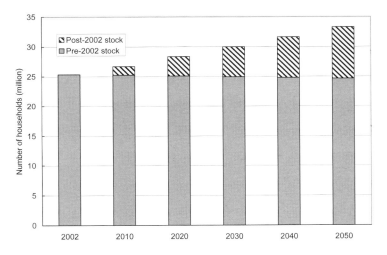

Figure 3 Households in existing stock built to 2002 standards or better

To give the Government more powers to control the energy efficiency of the existing stock through Building Regulations, the Building Act (which gives the powers to make Building Regulations) was amended in 2004 by the Sustainable and Secure Buildings Act.

Controlled fittings and services, and thermal elements

The 2002 amendment to Part L introduced provisions to control the installation of certain 'fittings and services' in existing

---

[5] House of Lords Select Committee on Science and Technology Second Report, *www.parliament.the-stationery-office.co.uk*, July 2005

buildings – including windows and boilers in dwellings, and windows, boilers, air-conditioning plant and lighting in other buildings.

The new review investigated how other types of work on existing buildings could be controlled, including work on 'thermal elements' – that is, the opaque external fabric of a building separating conditioned from unconditioned spaces (see Definitions in Chapter 6).

Consequential improvements

One proposal in the consultation paper called for cost-effective 'consequential improvements' to be made to parts of existing buildings not affected by new building work. An example of a consequential improvement in a dwelling would be fitting cavity wall insulation or extra loft insulation when putting in a new central heating system. An example in a non-domestic building would be upgrading old heating, cooling or ventilation plant when building an extension.

The original proposal was for the requirement to apply to all buildings, but as a result of the consultation the requirement has been restricted to buildings with a floor area greater than 1000 m^2.

Review of sustainable buildings

The Government's view was that it would be premature to include a requirement for consequential improvements for dwellings in Part L. Instead, it announced that ODPM would carry out a review, along with the Treasury, DTI and Defra, to identify measures to increase the overall sustainability of existing buildings, not just their energy efficiency.

> ODPM will review measures to increase overall sustainability of existing buildings, not just energy efficiency

The review would consider Building Regulations alongside other issues, such as the role of incentives, voluntary initiatives and the Home Information Pack (the 'sellers pack', which is due to come into force in June 2007). The review is due to be completed in the summer of 2006.

> **Code for Sustainable Homes**
> The Building Regulations set minimum standards, but the Government is already encouraging builders to do more by introducing a voluntary Code for Sustainable Homes.
>
> The Code is a voluntary scheme developed with industry which sets environmental performance standards for homes that are higher than those stipulated by regulation. The Government wants it to become a national code for sustainable homes that all sectors of the building industry will subscribe to and consumers will demand.
>
> A consultation paper was published by ODPM in December 2005, which sets out performance-based standards for energy, waste, water and materials efficiency.
>
> The higher standards in the Code will form the basis of the next wave of improvements to the Building Regulations.

Measures for improving compliance

One of the Government's commitments in the Energy White Paper was to investigate with local authorities (and by inference Approved Inspectors) how compliance with Part L requirements could be cost-effectively improved. Surveys by BRE and others had shown that the as-built performance of buildings was not always satisfactory, due to a combination of poor understanding of Part L requirements, poor workmanship and local authorities' limited resources for enforcing compliance.

The 2006 revisions to Part L aim to improve compliance by:
- introducing simple checklists into the Approved Documents
- simplifying the technical guidance and making increased reference to 'second tier' documents containing approved industry guidance
- building the rules for compliance into the software of the National Calculation Methodology (see Chapter 5) so that they are not open to interpretation
- authorising a number of additional 'competent person' schemes, whose registered members are allowed to self-certify compliance with certain Part L requirements (for example energy calculations and pressure testing) – to reduce the burden on both building control bodies (local authorities or private approved inspectors) and the construction industry

The Government, working together with Building Control Bodies, also has underway the largest ever training and dissemination programme for new building regulations. This programme of seminars, regional roadshows and workshops targeted at building control surveyors in both the public and private sectors started in September 2005 and will include an e-learning pack for every building control surveyor.

The 2004 Sustainable and Secure Buildings Act will also help by allowing regulations to be made requiring the appointment of a single person to manage compliance with Building Regulations for the life of a building project.

Competent person schemes

Competent person schemes have been authorised to register installers doing the following types of work in Part L-related areas:
- installation of gas, oil or solid fuel combustion appliances, or associated heating or hot water system or controls, in dwellings or in non-domestic buildings with no more than three storeys (there are three separate schemes for gas, oil and solid fuel appliances)
- installation of heating or hot water service systems or associated controls in dwellings
- installation of heating, hot water service, mechanical ventilation or air conditioning systems or associated controls in buildings other than dwellings
- installation of air-conditioning or ventilation systems in existing dwellings which does not involve work on systems shared with other dwellings
- installation of commercial kitchen ventilation systems which does not involve work on systems shared with parts of the building occupied separately
- installation of lighting or electric heating systems or associated electrical controls
- installation of replacement windows, rooflights, roof windows or glazed (more than 50%) doors
- air pressure testing of buildings
- carbon dioxide emission rate calculations.

Installers registered with competent person schemes must self-certify that their work complies with all parts of the Building Regulations, not just with Part L.

The last two schemes differ from the others in not being concerned with the installation of products, but rather with air pressure testing and carbon dioxide emission rate calculations.

Whereas competent persons registered with installation schemes self-certify that their work complies with all Parts of the Building Regulations, including Part L, competent persons registered with the last two schemes self-certify compliance only with a specific requirement of the Regulations

> Installers registered with competent person schemes must self-certify that their work complies with all parts of the Building Regulations, not just Part L

– the first limiting the building's air leakage, and the second limiting the building's carbon dioxide emissions.

There are also other competent person schemes for plumbing and electrical work (Part P).

3 Energy Performance of Buildings Directive

A Government commitment in the Energy White Paper was to implement the EU's Energy Performance of Buildings Directive (EPBD), which was published on 4 January 2003.

The objective of the EPBD is to promote the introduction of cost-effective measures, including renewable energy systems, to improve the energy performance of new and existing buildings. It recognises that the largest potential for energy savings lies with the existing building stock.

> The largest potential for energy savings lies with the existing building stock

Requirements

The Directive's detailed requirements are contained in 16 articles, although the main requirements are in Articles 3 to 10 and 15 (see box on the next page).

In essence EU Member States must introduce laws from 4 January 2006 that:

- set minimum energy performance standards for new buildings and existing large (more than 1000 m²) buildings when they undergo major renovation
- require a feasibility study to be carried out for new large buildings into the potential for using renewable energy sources, combined heat and power systems (CHP), district heating and cooling, and heat pumps

- require an energy performance certificate to be made available whenever a building is constructed, sold or rented out, and in addition displayed in large 'public' buildings

15

- require regular inspections by independent experts of solid, liquid and (optionally) gas fuel boilers with an output of more than 20 kW (although with an option to provide advice instead); a one-off assessment of heating systems with boilers that are more than 15 years old; and regular inspections of air-conditioning systems with a cooling capacity of more than 12 kW

Energy Performance of Buildings Directive — Articles

Article 3   *Adoption of a methodology* — National Calculation Methodology: SAP, SBEM and commercial software

Article 4   *Setting of energy performance requirements* for ...

Article 5   ... *new buildings* — whole building targets; consider feasibility of alternative energy sources for large buildings

Article 6   ... *existing buildings* over 1000 m^2 undergoing major renovation

Article 7   *Energy performance certificate* — on construction, sale and rental, and for display in large public buildings

Article 8   *Inspection of boilers* over 20 kW

Article 9   *Inspection of air conditioning systems* over 12 kW

Article 10 *Independent experts*

Article 15 *Transposition* — by 4 January 2006, or of articles 7 to 9 by 4 January 2009 if shortage of independent experts

---

To facilitate comparisons between buildings, the energy performance figures quoted must be based on nationally agreed calculation methods to a framework specified in the Directive.

Also, certificates must include benchmarks and indicate ways of improving performance. The certificates are intended to raise awareness among people buying and renting buildings of the opportunities for saving energy. For example, they provide an incentive to landlords to invest in energy saving measures – at the moment, landlords have no such incentive because energy costs are borne by tenants.

The Directive requires that the energy certification of buildings and inspection of boiler and air conditioning plant should be carried out in an independent manner by qualified and/or accredited experts.

Member States may be given until 4 January 2009 to fully implement certification and inspection requirements if they can show that there is a shortage of accredited experts.

4 Implementing the changes to Part L

Domestic boiler amendment — April 2005

The Government began implementing its proposals for amending Part L with an amendment to Approved Document L1 in 2005 to raise the efficiency standards for new and replacement domestic boilers. The amendment required most gas boilers to be of the highest efficiency condensing type from April 2005 and most oil boilers to be condensing from April 2007.

Main amendment — April 2006

The remaining more wide-ranging amendments to Part L to raise standards generally were introduced in April 2006.

EPBD Articles 3 to 6

Part L 2006 of the Building Regulations implements Articles 3 to 6 of the EPBD. It does this by:
- setting whole building carbon dioxide emissions targets for new buildings
- setting traditional 'elemental' performance standards for alterations to existing buildings (with an option to use carbon dioxide targets where appropriate)
- specifying the National Calculation Methodology for calculating energy performance, based on
  o Standard Assessment Procedure (SAP 2005) for dwellings
  o Simplified Building Energy Model (SBEM) or approved commercial software for other buildings.

Low and zero carbon systems

Article 5 of the EPBD requires consideration to be given before construction starts to incorporating low and zero carbon (LZC) energy supply systems into buildings with a floor area greater than 1000 m².

> Before construction starts, consider incorporating low and zero carbon energy supply systems

Part L 2006 implements this requirement by including an 'LZC benchmark' provision of 10% in the Target Emission Rating (TER) for all sizes of buildings that are not dwellings (see Chapter 5).

What this means is that the limit for carbon dioxide emissions has been lowered by an amount that is 10% beyond the level that can readily be achieved using conventional building materials and services, to encourage builders to adopt LZC energy systems such as combined heat and power (CHP), solar panels, ground source heat pumps and biomass.

Since Building Regulations are not prescriptive, builders are free to adopt other measures – such as higher levels of insulation – to meet the 10% higher standard for carbon emissions, but they may find that using LZC sources is the easier and more cost-effective route.

To help builders, the ODPM has published a guide to LZC energy sources[6]. Annex 5 gives more information.

EPBD Articles 7 to 10

Articles 7 to 10 will be implemented in stages by 4 January 2009 using other regulations.

Certification of energy performance

Housing regulations, for example, will be used to implement energy certification (Article 7) for the marketed sale of houses – to coincide with the introduction of the Home Information Pack (HIP) in June 2007.

Figure 4 shows the first two pages of the Energy Performance Certificate, which will form part of the Home Condition Report in the HIP.

---

[6] *Low or zero carbon energy sources: strategic guide.* ODPM, 2006. Available from *www.odpm.gov.uk*

By January 2009, energy performance certificates will be required for all buildings on construction, sale and rent, and also for display in large public buildings.

The certificates could display the following energy ratings as appropriate:
- a design rating for buildings before construction
- an asset (as-built) rating for completed new buildings and existing buildings on sale or rent
- an operational rating based on metered energy consumption.

PART L EXPLAINED — THE BRE GUIDE

THIS IS AN EXAMPLE REPORT AND IS NOT BASED ON AN ACTUAL PROPERTY

Section H: Energy Performance Certificate SAP

100 Any Street,	Dwelling type:	Detached	Certificate number:	XXXX
Any Town,	Internal floor area:	XXXX	Date issued:	XXXX
Anywhere, AB1 CD2	Date of inspection:	XXXX	Name of inspector:	XXXX

This home's performance ratings

This home has been assessed using the UK's Standard Assessment Procedure (SAP) for dwellings. Its performance is rated in terms of the energy use per square metre of floor area, energy efficiency based on fuel costs and environmental impact based on carbon dioxide (CO_2) emissions.

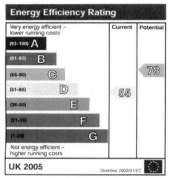

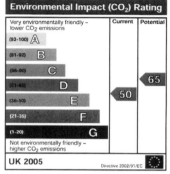

The energy efficiency rating is a measure of the overall efficiency of a home. The higher the rating, the more energy efficient the home is and the lower the fuel bills will be.

The environmental impact rating is a measure of a home's impact on the environment in terms of carbon dioxide emissions. The higher the rating, the less impact it has.

Estimated energy use, carbon dioxide (CO_2) emissions and fuel costs of this home

This table provides an indication of how much it will cost to provide lighting, heating and hot water to this home. This information has been provided for comparative purposes only. The fuel costs and carbon dioxide emissions are calculated based on a SAP assessment of the energy use. This makes standard assumptions about occupancy, heating patterns and geographical location.

The energy use includes the energy used in producing and delivering the fuels to this home. The fuel costs only take into account the cost of fuel and not any associated service, maintenance or safety inspection costs.

This certificate allows one home to be directly compared with another, but always check the date the certificate was issued. Since fuel prices can increase over time, an older certificate may underestimate the property's fuel costs.

	Current	Potential
Energy use	xxx kWh/m^2 per year	xxx kWh/m^2 per year
Carbon dioxide emissions	xx tonnes per year	xx tonnes per year
Lighting	£xxx per year	£xxx per year
Heating	£xxx per year	£xxx per year
Hot water	£xxx per year	£xxx per year

To see how this home can achieve its potential rating please go to page ii.

i

Figure 4 Energy performance certificate (front page)

IMPLEMENTING THE CHANGES TO PART L

Energy Performance Certificate
Report Section

Certificate number: XXXXXXXXXXXXXXXXXXXX
Date issued: XXXXXXXXXXXXXXXXX
Name of inspector: XXXXXXXXXXXXXXXXXXXX

Summary of this home's energy performance related features

The following is an assessment of the key individual elements that have an impact on this home's performance rating. Each element of this home is rated on the following scale: Very poor/ Poor/ Average/ Good/ Very good

Element	Description	Current performance
Main walls	Uninsulated cavity wall	Poor
Main roof	Pitched, 100mm loft insulation	Average
Main floor	Uninsulated solid concrete (assumed)	Average
Windows	Single glazed throughout	Very poor
Main heating	Mains gas back boiler	Average
Main heating controls	No controls	Very poor
Secondary heating	Flame effect fire	Very poor
Hot water	From main heating system; uninsulated cylinder	Very poor
Lighting	Low energy lighting in all fixed outlets	Very good
Current energy efficiency rating		**D 55**
Current environmental impact (CO_2) rating		**E 50**

Cost effective measures to improve this home's performance ratings

All the measures below are cost effective. The performance ratings after improvement listed below are cumulative, that is they assume the improvements have been installed in the order that they appear in the table.

Lower cost measures up to £500	Typical savings per year	Performance ratings after improvement	
		Energy efficiency	Environmental impact
1 Cavity wall insulation	£xx	D 65	D 56
2 Loft insulation top up to 250mm	£xx	C 68	D 57
3 Hot water cylinder and pipe work insulation	£xx	C 69	D 58
Sub-total	£xx		
Higher cost measures over £500			
4 Condensing boiler	£xx	C 75	D 63
5 Installation of a full heating controls package	£xx	C 78	D 65
Total	£xx		
Potential energy efficiency rating		**C 78**	
Potential environmental impact (CO_2) rating			**D 65**

Further measures to achieve even higher standards

The further measures listed below should be considered in addition to those already specified if aiming for the highest possible standards for this home.

6 Double glazing	£xx	C 80	C 67
7 Solar water heating	£xx	B 81	C 68
Enhanced energy efficiency rating		**B 81**	
Enhanced environmental impact (CO_2) rating			**C 68**

Improvements to the energy efficiency and environmental impact ratings will usually be in step with each other. However, they can sometimes diverge because reduced energy costs are not always accompanied by reduced carbon dioxide emissions.

Figure 4 Energy performance certificate (CONTINUED)

5 National Calculation Methodology

Carbon dioxide targets

From April 2006, the procedure for complying with Part L involves comparing the calculated carbon dioxide emissions from a proposed building with the calculated carbon dioxide emissions from a 'notional', or reference, building of the same size and shape, with a defined energy performance and operating under standardised conditions of occupancy and weather.

To meet Part L 2006 requirements, the amount of carbon dioxide emitted by the proposed building – the **Dwelling Emission Rate (DER)** for dwellings, or **Building Emission Rate (BER)** for other buildings – must be a specified percentage below the carbon dioxide emission rate of the notional building. The required percentage improvement or 'improvement factor' varies with the type of building – from 20% for gas-heated dwellings to 28% for air-conditioned buildings. This reduced carbon dioxide emission rate becomes the **Target Emission Rate (TER)**.

The energy performance of the notional building is close to that of a building with gas heating designed to meet Part L 2002 elemental standards for fabric and building services. So the improvement factor represents the improvement in standards since 2002.

This procedure for setting whole building standards for new buildings complies with the requirements of the EPBD, but also:
- gives designers flexibility to choose solutions that best meet their needs
- makes it easy to raise standards still further in future revisions of Part L.

Calculating carbon dioxide emissions

The National Calculation Methodology (NCM) is the system of rules specified in Part L for calculating the carbon dioxide emission rates for the proposed and the notional building. It specifies how the calculation is to be performed, the software tools to be used, and the standardised conditions for weather and occupancy, etc. The NCM therefore comprises the underlying method plus the standard data sets for different activity areas and for the construction and service elements.

> NCM is the system of rules for calculating the carbon dioxide emission rates for the proposed and notional buildings

The NCM can provide a pre-construction 'design' rating and a post-construction 'asset' rating. The asset rating is based on the as-built performance, and would, for example, include actual (rather than design) values of air leakage obtained from airtightness tests.

The NCM is described fully in Approved Documents L1A and L2A for new buildings, and in an associated second tier guidance document[7], which is available from *www.odpm.gov.uk*.

The actual calculations are performed using two tools based on European (CEN) standards: SAP 2005 and SBEM.

Standard Assessment Procedure (SAP)

SAP 2005, 'The Government's Standard Assessment Procedure for energy rating of dwellings', is specified in Part L 2006 as the method for calculating the amount of carbon dioxide emitted by dwellings with a floor area not greater than 450 m².

> SAP 2005 is the method for calculating the carbon dioxide emitted by dwellings with a floor area not greater than 450 m²

The carbon dioxide calculation, expressed in $kgCO_2/m^2/year$, takes account of space heating, water heating, ventilation and lighting, and emissions saved by energy generation technologies.

The full SAP 2005 specification[8] – comprising the text, tables and worksheet – can be downloaded from *www.bre.co.uk/sap2005* and there is more information about SAP in Annex 1 of this guide.

---

[7] *The National Calculation Methodology for determining the energy performance of buildings. Part 1: A guide to the application of the SBEM and other approved calculation tools for Building Regulations purposes*, ODPM, 2006

[8] *The Government's Standard Assessment Procedure for energy rating of dwellings, SAP 2005*

NATIONAL CALCULATION METHODOLOGY

Calculations should be performed using an approved software tool that implements the SAP specification. A list of approved SAP software can also be obtained from the above website.

SAP 2005 is a development of SAP 2001, which produced an energy cost rating (the SAP rating) and a carbon emissions rating (the Carbon Index). One way of complying with Part L 2002 was to show that the Carbon Index for a new dwelling was better than a defined threshold.

For the purposes of energy certification and showing compliance with Part L 2006 carbon targets, SAP 2005 can produce four ratings:
- the energy consumption per unit floor area
- an energy cost rating (the SAP rating)
- an Environmental Impact rating (based on carbon dioxide emissions), and
- the DER (Dwelling carbon dioxide Emission Rate).

A SAP 2005 report will also identify any "important design features" – for example better than average U-values, thermal bridging, air leakage and heating system efficiency, or the use of renewables – as listed in Appendix B of Approved Document L1A. This is intended to help building control bodies focus on the important details of a design when checking for compliance with the Regulations.

For a new dwelling, the single way now of complying with Part L requirements is to show that the DER is no worse than the TER (Target carbon dioxide Emission Rate).

> The single way of complying is to show that the DER is no worse than the TER

As explained earlier, the TER is derived from the carbon dioxide emission rate for the notional building. For gas-heated dwellings the TER is set at 80% of the carbon dioxide emission rate for the notional building. Attaining the target is more demanding for LPG, oil and electrically heated dwellings because of the higher 'carbon intensity' of these fuels (that is the amount of carbon dioxide they produce per kWh) compared with mains gas. However, this is partly mitigated by a 'fuel factor' that sets the target emissions about midway between the target for mains gas and what would apply taking no account of the differing carbon intensities of the fuels. This is explained in more detail in the section on design standards in Chapter 7.

PART L EXPLAINED — THE BRE GUIDE

Simplified Building Energy Model (SBEM)

SBEM is the default software tool specified in Part L 2006 for calculating the amount of carbon dioxide emitted from dwellings with a floor area greater than 450 m². and from all other buildings. SBEM is in two parts:
- SBEM the calculation engine,
- iSBEM the user interface.

They can both be downloaded from *www.ncm.bre.co.uk* together with a user guide[9].

> SBEM is specified in Part L for calculating the amount of carbon dioxide emitted from dwellings with a floor area greater than 450 m²

SBEM is applicable to most buildings. It calculates energy use and carbon dioxide emissions, based on a description of the building geometry, construction, use and HVAC and lighting equipment – all in accordance with emerging European (CEN) standards. SBEM makes use of standard data contained in associated databases.

There is further information about SBEM in Annex 2 of this guide.

Some software vendors are embedding SBEM within their own interfaces, and others will use accredited calculation tools – for example detailed simulation software – in place of SBEM.

---

[9] *Simplified Building Energy Model (SBEM) user manual and calculation tool*

6 Part L Regulations and approved guidance

Statutory Instruments

The Building Regulations 2000 were published as Statutory Instrument SI 2000/2531[10]. This has been amended a number of times since 2000 as changes have been made to various Parts of the Regulations. For example, Part L revisions were implemented in April 2002 by SI 2002/440 and in April 2006 by SI 2006/652.

Requirements in the Building Regulations are of both a technical and procedural nature. Regulation 4 is a key technical one, in that it requires 'building work', as defined in Regulation 3, to comply with the performance-based 'Requirements' listed in Schedule 1, under the 14 Parts A to P. The Part L performance-based Requirement is that **"reasonable provision shall be made for the conservation of fuel and power in buildings"**.

Other regulations have been added in 2006, including 17A, B and C, which make it a requirement that carbon dioxide emissions from whole buildings should meet specified targets.

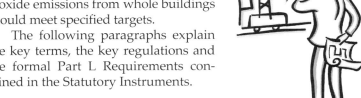

The following paragraphs explain the key terms, the key regulations and the formal Part L Requirements contained in the Statutory Instruments.

Definitions

The following are informal, short definitions and explanations of Building Regulations terms. Most are defined formally under Regulation 2 of the Building Regulations 2000 as amended or in the Approved Documents:

---

[10] *www.odpm.gov.uk/building-regulations* > Building Act 1984 and the Building Regulations > legislation

Building – any permanent or temporary building, including part of a building.

The types of building to which the energy efficiency provisions in the Regulations apply are identified in Regulation 9 and Schedule 2 of the Regulations. The sections below on 'Regulations' and on 'Exempt buildings and work' gives further details.

Dwelling – a self-contained unit designed to accommodate a single household. It includes a dwelling-house and a flat.

A building that is not a dwelling is generally referred to in this guide as a **'non-domestic building'** or as a **'building other than a dwelling'**.

An occupied building that is not a dwelling is essentially a non-domestic workplace.

A building containing living accommodation and also a small amount of space for commercial purposes – such as a workshop or an office – should be treated as a dwelling if the commercial part could revert to domestic use on change of ownership.

Where a dwelling is part of a larger building containing other types of accommodation, the non-dwelling parts should be treated as non-domestic buildings. The common areas of buildings containing multiple dwellings should also be treated as non-domestic buildings.

Mixed-use development – a building in which part may be used as a dwelling while another part has a non-domestic use – for example commercial or retail. The guidance in the Approved Documents for buildings other than dwellings apply to the non-domestic part.

Room for residential purposes – a room or suite of rooms that is used by one or more persons to live and sleep in. It includes a room in a hostel, a hotel, a boarding house, a hall of residence or a residential home, but does not include a room in a hospital or other similar establishment used for patient accommodation. A room for residential purposes is not a self-contained unit and is therefore not a dwelling. A building containing rooms for residential purposes is also not a dwelling.

House in multiple occupation (HMO) – a dwelling shared by several people who are not members of the same family but who share a kitchen or WC – for example a bedsit.

Conservatory – an extension to a building that is mainly glazed (at least three-quarters of the roof and half of the external walls), and is also thermally separated from the building. The thermal separation must be comparable to the rest of the external envelope of the building.

A conservatory with a floor area no greater than 30 m² is currently exempt from Part L of the Building Regulations. If a substantially glazed space is not a conservatory, or is a conservatory with a floor area greater than 30 m², then the space – whether part of a new building or an extension to an existing building – must comply with Part L requirements.

> A conservatory with a floor area no greater than 30 m² is currently exempt from Part L

Fixed building services – fixed systems, including controls, for heating, hot water, air conditioning or mechanical ventilation; and fixed internal or external lighting systems (but excluding emergency escape lighting or specialist process lighting).

Part L controlled service or fitting – a service or fitting upon which Part L imposes a requirement. A Part L controlled service is a fixed building service, and a Part L controlled fitting is a window, roof window, rooflight or door.

Thermal element – a wall, floor or roof (an opaque element of the building fabric that is not a fitting) which separates the internal conditioned space from the external environment, including from a space that is not conditioned such as an unheated garage, storage area or plant room.

Material alteration – work that could have a detrimental impact on the structure of a building, or on fire safety or access for the elderly and disabled.

Material change of use – work that involves converting a building or part of a building into a dwelling, a room for residential purposes, a hotel, a boarding house, an institution, a public building or a shop. It includes work to convert an exempt building into one that is no longer exempt.

Renovation – the provision of a new layer in the thermal element of a building, or the replacement of an existing layer. Decorative finishes are not thermal layers and are excluded.

Building work – work subject to Building Regulations control, as defined in Regulation 3. It includes:
- construction of a new building, and
- the following types of work in existing buildings:
  - construction of an extension
  - provision (including replacement), extension or alteration of a controlled service or fitting
  - provision (including replacement), renovation or retention of a thermal element
  - material change of use
  - material alteration
  - replacement or renovation of a thermal element
  - works associated with a change in a building's energy status
  - consequential improvements.

The last three involve compliance only with Part L requirements unless they are also material alterations.

Energy efficiency requirements – the requirements of Part L of Schedule 1 along with the new requirements in Regulations 4A, 17C and 17D (see below).

Change to a building's energy status – any change that results in a building becoming a building to which the energy efficiency requirements of these Regulations apply, where previously it was not.

Notifiable work – building work that must either be approved by a building control body (local authority or private approved inspector) or else self-certified by a competent person registered with a relevant authorised scheme. Some types of minor building work, although legally required to comply with the standards in the Building Regulations, are not notifiable.

Consequential improvement – a cost-effective improvement made to an existing building when other building work is carried out – for example installing cavity wall insulation, or topping up loft insulation when installing a new

> Consequential improvements to buildings with a floor area over 1000 m²

PART L REGULATIONS AND APPROVED GUIDANCE

heating system in a dwelling; or upgrading old heating, cooling or ventilation plant when extending a non-domestic building. The Regulations (Regulation 17D) call for consequential improvements to be made only to buildings with a floor area over 1000 m².

Regulations

A summary of the key regulations in the Building Regulations 2000 as amended, including all the regulations referred to in the Approved Documents, follows.

The regulations contain some requirements that are technical – saying how building work must be carried out, tested and commissioned – and others that are procedural – for example requiring full plans or a building notice to be submitted before starting building work, or a commissioning certificate to be issued on completion of work on building services.

Regulation 2 defines the key terms such as 'building' and 'dwelling'.

Regulation 3 identifies the types of building work that are controlled by Building Regulations.

Regulation 4 requires building work, as defined by Regulation 3, to comply with the performance 'Requirements' listed in Schedule 1, under the 14 Parts A to P. The Part L Requirements are reproduced later in this Chapter in the section 'Requirements in Schedule 1'.

Regulation 4A requires thermal elements to be reasonably energy efficient if they are renovated or replaced.

Regulation 4B requires that, where there is a change to the energy status of a building, any necessary work should be carried out to ensure that the building complies with the applicable requirements of Part L of Schedule 1. Such work is required to comply with other Parts of Schedule 1 only if it is also a material alteration.

Regulation 5 defines 'material change of use' and *Regulation 6* identifies the Parts of Schedule 1 that are applicable to a material change of use.

Regulation 7 requires building work to be carried out with adequate and proper materials and in a workmanlike manner.

Regulation 9 identifies the types of work and types of building – for example small extensions and detached buildings listed in Schedule 2 – that are exempt from certain requirements in the Regulations. For example conservatories with a floor area no greater than 30 m^2 are generally exempt, but glass must comply with Part N standards for safety glazing, and electrical installations must comply with Part P standards for electrical safety.

From April 2006, the exemptions in Schedule 2 no longer apply generally to the energy efficiency requirements in the Regulations, for which there are new provisions (see below).

Regulation 11 gives local authorities the power to dispense with or relax any requirement contained in the Regulations.

Regulation 12 requires a person who intends to carry out building work to give to the local authority a 'building notice' (at least two working days before starting work); or to deposit 'full plans' (for approval before starting work). (A person may alternatively use a private approved inspector as the building control body.)

Regulation 12 also states that a person registered with one of the competent person self-certification schemes listed in Schedule 2A or carrying out 'non-notifiable' minor work described in Schedule 2B is not required to issue a building notice or deposit full plans.

Regulations 13 and 14 give details of the information to be provided and other requirements when submitting a building notice or full plans.

Regulation 15 gives the requirements on when to notify the local authority before starting work, during construction, on completion, and before occupation.

Regulation 16 requires a SAP energy rating (see Chapter 5) to be prepared and fixed in a conspicuous place in a newly constructed dwelling before occupation. A copy of the SAP rating must also be given to the local authority.

Regulation 16A sets out the arrangements for a Competent Person to self-certify compliance with the Building Regulations by issuing a 'compliance certificate' to a building occupier and notifying the local authority within 30 days of completing work.

Regulation 17 sets out the arrangements for a local authority to issue a Building Regulations 'completion certificate' to an occupier for notifiable work that has not been carried out by a Competent Person.

Regulation 17A requires a methodology to be approved for calculating carbon dioxide emissions from buildings.

Regulation 17B requires carbon dioxide emission targets to be set for new buildings.

Regulation 17C requires new buildings to have carbon dioxide emission rates that are no worse than approved targets, calculated in accordance with an approved methodology.

Regulation 17D requires practicable and cost-effective 'consequential improvements' to be made to existing buildings over 1000 m^2 when they are extended, fitted with new building services, or the building services are extended.

Regulation 17E states that, for the purposes of Part L 'building' means the building as a whole or parts of it that have been designed or altered to be used separately.

Regulation 20B requires a pressure test notice to be given to the local authority confirming that new buildings have been pressure tested in accordance with an approved procedure. The local authority is authorised to accept, as evidence that an approved procedure has been followed, a certificate from a person Registered with the British Institute of Non-destructive Testing.

Regulation 20C requires a commissioning notice to be given to the local authority following the installation or extension of fixed building services confirming that the fixed building services have been commissioned in accordance with an approved procedure.

Regulation 20D requires a carbon dioxide emission rate certificate to be given to the local authority specifying the target emission rate (TER) for the building and the calculated emission rate for the building as constructed. The local authority is authorised to accept, as evidence that the TER will not be exceeded, a certificate from a person registered with FAERO Ltd or BRE Certification Ltd.

Exempt buildings and work

Schedule 2 of the Building Regulations used to list buildings and work which were exempt from the Building Regulations, including Part L. From April 2006, the only building work in Schedule 2 that is exempt from the new energy efficiency requirements of the Regulations (Regulations 4A, 17C and 17D, and Part L of Schedule 1) is the extension of a building by the addition at ground level of a conservatory or enclosed porch with a floor area not exceeding 30 m². Other buildings and work that are exempt from the energy efficiency requirements are now identified in Regulation 9 paragraphs (4) and (5) as being:

- buildings that do not use energy to condition the indoor climate (that is, buildings where energy is consumed directly by a commercial or industrial process)
- protected buildings – listed buildings, buildings in conservation areas, and ancient monuments – where compliance with the energy efficiency requirements would unacceptably alter their character or appearance
- buildings used as places of worship
- temporary buildings with a planned life of no more than two years
- industrial sites, workshops and non-residential agricultural buildings with low energy demand
- stand-alone buildings other than dwellings, with a floor area of less than 50 m².

The changes have been introduced largely to conform with new requirements in the EPBD.

Transitional arrangements

The revised Part L 2006 Regulations came into effect on 6 April 2006. Transitional arrangements – whether using a local authority or a private Approved Inspector as the building control body –

allow some work carried out after 6 April to comply with the Part L 2002 Regulations if:
- work was started before 6 April 2006 (usually marked by work on foundations or drains)
- for works which do not require full plans approval, a contract was entered into before 6 April 2006, provided that the work is started before 1 October 2006. This is intended to help smaller builders and householders who may not be fully aware of the technical changes, particularly in view of the short period between the announcement of the transitional provisions and the coming into force of the amendments

> For plans approved by April 2006, builders have until April 2007 to start work if they are to be allowed to comply with the 2002 Regulations

- plans were approved by a local authority before 6 April 2006, provided that work is started before 1 April 2007. This represents a tightening up of the usual transitional arrangements in that builders will now have only one year – rather than three years as before – to start work once plans have been approved if they are to be allowed to comply with the 2002 Regulations.

Special arrangements apply to modular and portable buildings that are not dwellings. Reasonable provision for energy efficiency would be to follow the guidance in *Energy performance standards for modular and portable buildings*[11] if:
- more than 70% of the external envelope is to be created from sub-assembles manufactured before 6 April 2006, and which are obtained from a centrally held stock or from the dismantling of buildings on other premises; or
- the intended life of the building is less than two years.

There are also transitional arrangements that apply to mandatory pressure testing of buildings in the period up to 31 October 2007 (see later).

Requirements in Schedule 1
Regulation 4 states that building work shall comply with the performance-based requirements in Schedule 1 of the Building

---
[11] *Energy performance standards for modular and portable buildings*, MPBA, 2006

Regulations. The current Requirements for Part L *Conservation of fuel and power* are:

> **L1.** Reasonable provision shall be made for the conservation of fuel and power in buildings by:
> a. limiting heat gains and losses:
>   i. through thermal elements and other parts of the building fabric; and
>   ii. from pipes, ducts and vessels used for space heating, space cooling and hot water storage;
> b. providing energy efficient and properly commissioned fixed building services with effective controls;
> c. providing to the owner sufficient information about the building, the fixed building services and their maintenance requirements so that the building can be operated in such a manner as to use no more fuel and power than is reasonable in the circumstances.

What constitutes reasonable provision will depend on the type of work (for example new or renovation) and the type of building (for example modern or historic). The Approved Documents (see below) give guidance.

Building work must satisfy all the requirements set out in Schedule 1 of the Building Regulations. Also relevant when considering the incorporation of energy efficiency measures in buildings are the requirements in Part C (Site preparation and resistance to contaminants and moisture), Part E (Resistance to the passage of sound), Part F (Ventilation), Part J (Combustion appliances and fuel storage systems) and Part P (Electrical safety).

Guidance on complying with the Regulations

Approved Documents

Approved Documents are intended to show one or more ways of carrying out common types of building work. However, there may well be alternative ways of achieving compliance with the requirements in the Regulations.

ODPM has published four Part L Approved Documents:
- **ADL1A:** Conservation of fuel and power in new dwellings
- **ADL1B:** Conservation of fuel and power in existing dwellings

- **ADL2A:** Conservation of fuel and power in new buildings other than dwellings
- **ADL2B:** Conservation of fuel and power in existing buildings other than dwellings.

Second tier guidance

The new Approved Documents give some technical detail, but also refer for further detail to approved 'second tier' guidance documents prepared by government and industry bodies, as listed in Table 1.

Changes to Part L introduced in 2006

The main changes introduced in 2006 are:
- There are now **four Approved Documents** instead of two
- More use is made of approved, **second tier technical reference documents**
- In each of the 2002 Approved Documents, there were three methods for showing that new buildings complied with Part L requirements: in ADL1 the elemental, target U-value and carbon index methods; and in ADL2 the elemental, whole building and carbon emission calculation methods. These have been replaced by **one method**: to show that the proposed building's carbon dioxide emissions rates (DER/BER) are no worse than the target emissions rate (TER) derived from a notional building of the same size and shape (new Regulations 17A, B and C). Limits are, however, placed on U-values and the amount of solar gain in non-conditioned buildings

  > The DER/BER must be no worse than the TER

- The **elemental approach** is retained for existing buildings, although for greater flexibility there is the option to use a whole building calculation model (such as SAP 2005 or SBEM) for material changes of use and extensions (to show that carbon dioxide emissions from the converted or extended building are no worse than would have been achieved by the elemental improvements required)
- Carbon dioxide emissions standards for **new buildings** are tightened – by around 20% on average for dwellings, 23% for naturally ventilated buildings, and up to 28% for air-conditioned buildings

  > Carbon dioxide emissions standards for new buildings are tightened by up to 28%

Table 1 Second tier guidance documents*

A practical guide to ductwork leakage testing. HVCA, 2000, DW143
Airtightness in commercial and public buildings. BRE Report BR 448, 2002
Assessing the effects of thermal bridging at junctions and around openings. BRE, 2006, IP1/06
Building energy metering. CIBSE, 2006, TM39
Building log book toolkit. CIBSE, 2003, TM31
Building regulations and historic buildings. English Heritage, 2002, Interim guidance note
CIBSE standard tests for the assessment of building services design software. CIBSE, 2004, TM33
Climate change and the indoor environment: impacts and adaptation. CIBSE, 2005, TM36
CO_2 emission factors for policy analysis, July 2005 www.bre.co.uk/filelibrary/co2emissionfigures2001.pdf
Commissioning management. CIBSE, 2003, Commissioning Code M
Conventions for U-value calculations. BRE, 2006, BR 443
Delivered energy emission factors for 2003, December 2005 www.bre.co.uk/filelibrary/2003EmissionFactorUpdate.pdf
Design for improved solar shading control. CIBSE, 2006, TM37
Domestic heating compliance guide. ODPM/NBS, 2006
Energy efficient ventilation in dwellings – a guide for specifiers on requirements and options for ventilation. EST, 2006, GPG268
Energy performance standards for modular and portable buildings. MPBA, 2006
Environmental design. CIBSE, 2006, Guide A
Guidance for design of metal roofing and cladding to comply with Approved Document L2 (2006). MCRMA, TP17
Guidelines for environmental design in schools. DfES, 2003, BB 87
HVAC guidance for achieving compliance with Part L of the Building Regulations. TIMSA, 2006
Lighting for buildings. Code of practice for daylighting. BSI, 1992, BS 8206-2
Limiting thermal bridging and air leakage: robust construction details for dwellings and similar buildings. ODPM, 2002
Low energy domestic lighting. EST, 2006, GIL20
Low or zero carbon energy sources: strategic guide. ODPM, 2006
Measuring air permeability of building envelopes. ATTMA, 2006, Technical Standard 1

PART L REGULATIONS AND APPROVED GUIDANCE

Table 1 Second tier guidance documents (CONTINUED)

Natural ventilation in non-domestic buildings. CIBSE, 2005, AM10
Non-domestic calculation methodology for Part L. ODPM, 2006
Non-domestic heating, cooling and ventilation compliance guide. ODPM, 2006
Reducing overheating — a designer's guide. EST, 2005, CE129
SBEM user manual and calculation tool. BRE, 2006, www.ncm.bre.co.uk
Selecting lighting controls. BRE, 2006, Digest 498
Solar shading of buildings. BRE, 1999, BR 364
Specification for sheet metal ductwork. HVCA, 1998, DW144
The Government's Standard Assessment Procedure for energy rating of dwellings. BRE/Defra, 2005
The National Calculation Methodology for determining the energy performance of buildings. Part 1: A guide to the application of the SBEM and other approved calculation tools for Building Regulations purposes. ODPM, 2006
Thermal assessment of window assemblies, curtain walling and non-traditional building envelopes. CWCT, 2006
Thermal insulation: avoiding risks. BRE, 2001, BR 262
Use of rooflights to satisfy the 2002 Building Regulations for the conservation of fuel and power. NARM, 2002
Ventilation of school buildings. DfES, 2005, BB101
Windows for new and existing housing. EST, 2006, CE66

* www.odpm.gov.uk/pub/12/CircularLetter30March2006_id1165012.pdf

- Standards for **existing buildings and extensions** have also been raised, except in the case of replacement windows for which the standards are generally unchanged (although standards have been raised for new windows in extensions)
- There is now a requirement to **pressure test** a sample of new dwellings as well as all new non-domestic buildings over 500 m^2 (new Regulation 20B). For small developments of dwellings and small non-domestic buildings, the need for testing can be avoided by assuming a high default value of air permeability for the calculation of DER or BER
- More **work on existing buildings** is captured. Part L 2002 already controlled the provision and extension of controlled fittings and services, and the provision of new thermal

elements of the building fabric. Part L 2006 now also requires:
- o cost-effective improvements to the energy efficiency of thermal elements when they are renovated (new Regulation 4A)
- o cost-effective consequential improvements to buildings over 1000 m^2 when subject to major works (new Regulation 17D)

- Part L now contains a requirement to provide evidence of commissioning of **building services** (new Regulation 20C) in new and existing buildings
- There is a **checklist** for builders and building control bodies to help in assessing the compliance of new buildings
- To help builders and building control bodies at the design stage of dwellings, **SAP 2005** will issue a warning if the energy performance of certain important design features is worse than a defined threshold
- There are new competent person **self-certification schemes** covering building services, air-pressure testing and carbon dioxide emission calculations
- Certain **minor works** on heating, ventilation and lighting systems no longer need to be notified to a building control body (although they must still comply with Part L technical requirements)
- There have been significant changes to **exempt buildings and work**.

7 Construction of new buildings

This Chapter presents an overview of the guidance in Approved Documents ADL1A and ADL2A on the construction of new dwellings and new non-domestic buildings.

It is not intended to replicate the guidance in the Approved Documents, but to highlight the key requirements. It provides further explanation where desirable, and explains the differences between the requirements for dwellings and other buildings.

The Approved Documents should be consulted for the full details, and, to help the reader to link between the documents, this section adopts the same main headings.

General guidance

Types of work covered
The 'Definitions' section in Chapter 6 explains when buildings should be treated as dwellings and when as non-domestic buildings, and describes the types of building work covered by Part L.

Approved Document ADL1A does not cover the conversion of buildings into dwellings – this is covered by ADL1B.

Demonstrating compliance
The Approved Documents lay down five criteria that need to be met to comply with the Regulations:

- **Criterion 1:** The predicted carbon dioxide emission rate for the proposed dwelling/building (DER or BER must not exceed the target emission rate (TER) specified in Approved Documents ADL1A and ADL2A

41

- **Criterion 2:** The performance of the building fabric and the heating, hot water and fixed lighting systems/building services systems must be no worse than the design limits specified in Approved Documents ADL1A and ADL2A
- **Criterion 3:** The dwelling/spaces without air conditioning must have appropriate passive control measures to limit the effects of solar gain
- **Criterion 4:** The performance of the dwelling/building as built must not exceed the Target carbon dioxide Emission Rate (TER)
- **Criterion 5:** The necessary information must be provided to permit energy efficient operation of the dwelling/building

> The Approved Documents lay down five criteria that need to be met to comply with the Regulations

A checklist in Appendix A of the Approved Documents is designed to help builders, developers and building control bodies confirm that the criteria have been met. The checklist, which can be downloaded from the ODPM website, should be used at both the design stage and the as-built stage. For each check it prompts for the evidence that needs to be provided and by whom.

Design standards

Criterion 1 — Predicted carbon dioxide emission rate to be no greater than target

Proposed buildings must be designed so that their design carbon dioxide emission rate is not greater than a Target carbon dioxide Emission Rate (TER) expressed in kg of carbon dioxide per square metre of floor area per year ($kgCO_2/m^2 year$).

The design rate for dwellings is the Dwelling carbon dioxide Emission Rate (DER); and for other buildings it is the Building carbon dioxide Emission Rate (BER).

The method to be used for calculating the design and target carbon dioxide emission rates is SAP 2005 for dwellings (with a floor area no greater than 450 m^2); and SBEM, or equivalent approved software, for other buildings.

SAP 2005 takes account of carbon dioxide emissions arising from heating, hot water and internal fixed lighting services; SBEM also takes account of carbon dioxide emissions arising from mechanical cooling and ventilation services.

CONSTRUCTION OF NEW BUILDINGS

SAP caters for only one type of building – dwellings; SBEM caters for dwellings over 450 m² and non-domestic buildings with a number of different activity areas – office accommodation, sports facilities, etc.

Where a building contains more than one dwelling (for example a terrace of houses or a block of flats), an average TER can be calculated for the whole building.

The DER and BER must be calculated both before and after construction to check that the as-built carbon dioxide emission rate does not exceed the TER.

Calculations should be undertaken by a person who is registered with either FAERO Ltd or BRE Certification Ltd, who is deemed to be competent to issue a certificate in respect of the calculation of carbon dioxide emission rates[12].

Calculating the Target carbon dioxide Emission Rate (TER)
The TER for the proposed building is calculated in two stages:

Stage 1: First, using SAP 2005 or SBEM as appropriate, the carbon dioxide emission rate is calculated for a notional (reference) building ($C_{notional}$) of the same size and shape as the proposed building.

Reference values for the fabric and fixed building services for the notional dwelling are set out in Appendix R of SAP 2005; and for the notional non-domestic building in SBEM[13]. These reference values typically correspond to those in a Part L 2002 building.

The reference value for air permeability[14] is 10 m³/hm² at 50 Pa. Until 31 October 2007 only, a higher measured value may be substituted if defined improvements are made after a failed mandatory pressure test[15]. ADL1A and ADL2A give the details.

---

[12] *www.faero.org.uk* and *www.bre.co.uk/energyrating*

[13] SBEM is available from *www.odpm.gov.uk* and *www.ncm.bre.co.uk*

[14] Air permeability (airtightness) is the air leakage rate per unit area of the building envelope at a reference internal-to-external pressure difference of 50 Pa, measured in units of cubic metres per square metre per hour. The air *permeability* of a building is measured by pressure testing

[15] For small developments with no more than two dwellings, or other buildings with a floor area below 500 m², pressure testing of a completed building is not mandatory if a high design air permeability of 15 m³/m²h at 50 Pa is assumed when calculating the DER or BER. If pressure testing is carried out, the measured value of air permeability is used in the final calculation of the DER or BER

43

ADL2A gives other required characteristics of the notional non-domestic building, which need to be known if using methods other than SBEM to calculate the TER (although SBEM will be suitable for many non-domestic buildings covered by this guide).

The notional non-domestic building, as well as being the same size and shape as the proposed building and using the specified reference values for fabric and services, should:
- have the same area of vehicle access doors and display windows as the proposed building
- exclude services not covered by Part L (for example emergency escape lighting and specialist process lighting)
- have the same activity areas and class of building services (taken from SBEM) as the proposed building
- have the reference occupancy times and environmental conditions for activity areas defined in SBEM
- be subject to a climate defined by the CIBSE Test Reference Year for the location of the proposed building
- use gas if available, or else oil, as the heating fuel, and use grid mains electricity for all other building services. Table 2 in ADL2A gives the carbon dioxide emission factors ($kgCO_2/kWh$) for the fuels

Stage 2: The TER is determined by reducing the notional carbon dioxide emission rate, $C_{notional}$, as follows:

For dwellings:

$$C_{notional} = C_H + C_L$$

where C_H is the carbon dioxide arising from the heating and hot water services and C_L is the carbon dioxide arising from the internal fixed lighting. Then:

$$TER = (C_H \times \text{fuel factor} + C_L) \times (1 - \text{improvement factor})$$

where the fuel factor varies from 1.00 for natural gas and renewables, to 1.17 for oil and 1.47 for grid electricity (see Table 2, and the improvement factor is 0.2 (that is 20%) for Part L 2006

The improvement factor represents the main reduction in carbon dioxide emissions called for by Part L 2006. In future revisions of Part L, it is the value that could be adjusted to reduce carbon dioxide emissions further.

CONSTRUCTION OF NEW BUILDINGS

The purpose of the fuel factor is to relax the TER for buildings in parts of the country where natural gas is not available, so that standards for buildings in these areas are not unreasonably high. Table 2 (which is based on Table 1 in ADL1A and Table 2 in ADL2A) gives the value of the fuel factor for different fuels, and shows also how the amount of carbon dioxide produced by the different fuels varies. The fuel factor adjustment does not compensate entirely for the extra carbon dioxide produced by oil, coal and electricity, and buildings that use these fuels will therefore need to have a better energy performance than buildings that use gas.

Table 2 Carbon dioxide emissions from different fuels

Fuel	CO_2 emissions $kgCO_2$/kWh	CO_2 emissions compared with natural gas	Fuel factor
Natural gas	0.194	1.00	1.00
LPG	0.234	1.21	1.10
Oil	0.265	1.37	1.17
Coal	0.291	1.50	1.28
Bio-fuels (bio-gas, wood pellets, etc)	0.025	0.13	1.00
Grid supplied electricity	0.422	2.18	1.47
Grid displaced electricity*	0.568		

*See Annex 5

For non-domestic buildings:

TER = $C_{notional}$ × (1 − improvement factor) × (1 − LZC benchmark)

The LZC benchmark is a further reduction in the target carbon dioxide emission rate – beyond what could readily be achieved by improvements in the energy performance of the building fabric and services – which has been introduced to encourage the use of low and zero carbon energy sources, such as solar panels and bio-fuels (see Annex 5).

The required Part L 2006 improvement factor and LZC benchmark for non-domestic buildings with heating, mechanical ventilation and air conditioning are shown in Table 3.

The overall reduction in carbon dioxide emissions (compared with the notional Part L 2002 building) required by Part L 2006 for non-domestic buildings varies from 23.5% to 28%.

Table 3 Required improvement in carbon dioxide emissions for non-domestic buildings

Type of building service	Improvement factor	LZC benchmark	Overall improvement
Heated and naturally ventilated	0.15	0.10	0.235
Heated and mechanically ventilated	0.20	0.10	0.280
Air-conditioned	0.20	0.10	0.280

Calculating the Dwelling/Building carbon dioxide Emission Rate (DER/BER)

The DER/BER of the actual building must not be greater than the TER. Two calculations must be performed using the same tool as for the TER:

Calculation 1: A preliminary calculation as part of the design submission, based on plans and specifications.

For dwellings, SAP 2005 will produce a report identifying the features of the design that are critical to the dwelling achieving the TER (see box opposite). Building control bodies will use the report during the construction phase to help them focus on the important features of the dwelling.

Calculation 2: A final calculation to demonstrate that the DER or BER of the actual building is no greater than the TER. The calculation must be based on the building as constructed, incorporating:
- any changes that have been made during construction
- the air permeability where measured (see below)
- for non-domestic buildings, the ductwork leakage and fan performance as commissioned

Other considerations when calculating the DER or BER are:

CONSTRUCTION OF NEW BUILDINGS

Pressure testing. The maximum design air permeability allowed by Part L when doing the first calculation of the DER or BER for the proposed building is 10 m³/hm² at 50 Pa. Following a pressure test of a completed building, the actual measurement is substituted for the second calculation.

Important design features for dwellings
If these threshold performance values are exceeded SAP 2005 will warn of the possibility of failing to comply:
- Wall U-value less than 0.28 W/m²K
- Floor U-value less than 0.20 W/m²K
- Roof U-value less than 0.15 W/m²K
- Window or door U-value less than 1.8 W/m²K
- Thermal bridging not greater than the default value for approved details
- Design air permeability less than 7 m³/hm² at 50 Pa
- Main heating system efficiency more than 4 percentage points better than that recommended for its type in the *Domestic heating compliance guide* (see Annex 6)
- The use of any low carbon or renewable energy technology such as:
  ○ bio-fuel used for the main heating system (including multi-fuel appliances)
  ○ CHP or community heating
  ○ heat pumps
  ○ a solar panel
  ○ a photovoltaic array
  ○ any item involving the application of SAP 2005 Appendix Q, which provides a method to allow for the benefits of new energy-saving technologies that are not included in the published SAP specification
---

For small developments, with no more than two dwellings, or other buildings with a floor area below 500 m², pressure testing is not mandatory if a high design air permeability of 15 m³/hm² at 50 Pa is assumed when calculating the DER or BER. This will require the building fabric and services to have a better energy performance than would

> For small developments or other buildings, pressure testing is not mandatory if a high design air permeability of 15 m³/hm² at 50 Pa is assumed

47

- **Secondary heating in dwellings.** For dwellings, SAP 2005 assumes for the calculation of the DER that a certain proportion of the space heat demand will be met by a secondary heating appliance. If no appliance is specified, SAP will assume that the secondary heating is provided by an electric room heater, or if there is a chimney or flue by either a low efficiency decorative fuel effect gas fire or an open fire. ADL1A gives further details.

- **Lighting in dwellings.** Part L 2006 requires the minimum number of fixed low energy light fittings to be one per 25 m^2 of floor area or part thereof, or one per four fixed light fittings, whichever is the greater.

  Builders are encouraged to install more low energy fittings than the minimum, but SAP 2005 calculates DER assuming 30% low energy lighting. This means that increasing the actual amount of low energy lighting in a dwelling will not reduce the DER, and so cannot be traded off against the energy performance required of the building fabric or other services.

- **Carbon dioxide emission factors.** ADL2A gives details of the carbon dioxide emission factors to be assumed for calculating the BER when using multi-fuel appliances (see Table 2). These figures are built in to the SBEM calculation tool.

  ADL2A also gives some guidance on how to estimate the carbon dioxide emission factors when thermal energy is supplied from a district heating or cooling system. The BER submission has to be accompanied by a report, signed by a suitably qualified person, explaining how the emission factors were derived.

- **Management features in non-domestic buildings.** In practice, energy efficiency can be improved by certain 'management features'. ADL2A allows the BER to be reduced by up to 5% if the actual building has power factor correction or automatic monitoring and targeting with alarms for out-of-range values.

CONSTRUCTION OF NEW BUILDINGS

- **Low and zero carbon energy sources.** ADL2A explains how low and zero carbon energy supply systems can, in appropriate circumstances, make substantial and cost-effective contributions to achieving TERs. ODPM has published a guide to the use of LZC sources on its website[16], which is summarised in Annex 5 of this guide.

  > ADL2A explains how low and zero carbon energy supply systems can make substantial contributions to achieving TERs

  LZC energy sources include:
  - solar hot water
  - photovoltaic power
  - bio-fuels – for example wood fuels and oil blends
  - combined heat and power at the building or community level
  - heat pumps.

For example, Table 2 gives a carbon dioxide emission factor of 0.568 kgCO$_2$/kWh for grid displaced electricity – generated by building-integrated power sources such as photovoltaic panels and combined heat and power. The associated carbon dioxide emissions are deducted from the total carbon dioxide emissions for the building to determine the BER, although any fuel used to power the CHP engine, for example, must be included in the calculation.

Criterion 2 – Performance to be within design limits

The Approved Documents set minimum values for the energy performance of elements of both the building fabric (for example maximum U-values for floors, walls and roofs) and the building services (for example minimum seasonal boiler efficiency).

Design flexibility is deliberately limited in this way for two reasons: first to avoid technical problems such as condensation; and second, to discourage excessive and inappropriate trade-off that might result, for example, in buildings being designed with poor insulation in combination with renewable energy systems that have an uncertain service life.

> Designing a building using only the minimum energy performance values will not be enough to meet the TER

---

[16] *Low or zero carbon energy sources: strategic guide*, ODPM, 2006

Designing a building using only these minimum energy performance values will not be enough to meet the TER.

Design limits for U-values

Approved Document ADL1A gives maximum allowed U-values for envelope elements, both area-weighted averages for a dwelling and worst values for individual elements, as shown in Table 4.

ADL2A sets additional values for pedestrian doors, vehicle access doors and similar large doors, high usage entrance doors and roof ventilators.

For the most part the new *limiting* area-weighted U-values in Part L 2006 have been set at the elemental standards in Part L 2002.

Other particular requirements are:
- **Display windows.** Display windows (normally below a height of 3 m) and similar glazing are not required to meet the limits in Table 4.
- **Determining U-values.** U-values should be calculated using the methods set out in the 2006 edition of BR 443, *Conventions for U-value calculations*[17].

    SAP 2005 Table 6e gives values for different window configurations that may be used in the absence of calculated or measured values.
- **High internal gains.** In non-domestic buildings with high internal gains, the maximum area-weighted average U-value for windows, doors and rooflights can be increased to 2.7 W/m²K if it can be shown that this will reduce carbon dioxide emissions.

Design limit for air permeability

The maximum design air permeability allowed by Part L when doing the first calculation of the DER or BER for a proposed building is 10 m3/hm2 at 50 Pa.

However, a lower value of air permeability (a less leaky building) may be needed to achieve the TER, and is technically desirable in buildings with mechanical ventilation, heat recovery and air conditioning.

---

[17] *Conventions for U-value calculations*, BR 443, BRE, 2006 edition.

CONSTRUCTION OF NEW BUILDINGS

Table 4 Limiting standards (U-values in W/m²K) for new buildings

Element	Area-weighted average U-value	Individual element U-value
Wall in dwelling and non-domestic building	0.35	0.70
Floor in dwelling and non-domestic building	0.25	0.70
Roof in dwelling and non-domestic building	0.25	0.35
Window, roof window and rooflight in dwelling and non-domestic building	2.2	3.3
Door in dwelling	2.2	3.3
Pedestrian door in non-domestic building	2.2	3.0
Vehicle access and similar large door in non-domestic building	1.5	4.0
High usage entrance door in non-domestic building	6.0	6.0
Roof ventilator (including smoke vent) in non-domestic building	6.0	6.0

Guidance on making buildings airtight is given in *Limiting thermal bridging and air leakage: robust construction details for dwellings and similar buildings*[18] and *Airtightness in commercial and public buildings*[19].

Design limits for fixed building services and controls in dwellings

ADL1A contains the following minimum design performance requirements for fixed building services and controls in dwellings:

---

[18] *Limiting thermal bridging and air leakage: robust construction details for dwellings and similar buildings*, Amendment 1, ODPM, 2006

[19] *Airtightness in commercial and public buildings*, BRE Report BR 448, 2002

PART L EXPLAINED — THE BRE GUIDE

- **Heating and hot water system appliances and controls:** Compliance with the *Domestic heating compliance guide*[20].

- **Mechanical ventilation, including air permeability standards for different ventilation strategies:** Compliance with *Energy efficient ventilation in housing: a guide for specifiers*[21]. In addition ADL1A sets the following design limits:
    o maximum specific fan power of 0.8 W/litre s for systems with continuous supply only and continuous extract only, and 2.0 W/litre s for balanced systems
    o minimum design limit of 66% for heat recovery efficiency.

- **Mechanical cooling:** Fixed air conditioners to have an energy classification equal to or better than Class C in Schedule 3 of the labelling scheme adopted under the Energy Information (Household Air Conditioners) (No 2) Regulations 2005[22]. SAP 2005 does not calculate energy use for cooling, but does check for high risk of overheating so that the need for cooling will be minimised.

- **Insulation of pipes, ducts and vessels:** Compliance with the *Domestic heating compliance guide*[20].

- **Fixed internal lighting:** Minimum number of fixed low energy light fittings (lamp luminous efficacy greater than 40 lumens per circuit-watt) to be either one per 25 m^2 of floor area or part thereof, or one per four fixed light fittings, whichever is the greater. *Low energy domestic lighting*[23] gives guidance on suitable locations.

- **Fixed external lighting:** Either low energy light fittings; or lamp capacity not to exceed 150 W, with automatic switch-off when not required (that is, when sufficient daylight, or people not detected).

---

[20] *Domestic heating compliance guide*, ODPM, 2006

[21] *Energy efficient ventilation in housing: a guide for specifiers on requirements and options for ventilation*, EST, GPG268, 2006

[22] Statutory Instrument 2005 No 1726

[23] *Low energy domestic lighting*, EST, GIL20, 2006

CONSTRUCTION OF NEW BUILDINGS

Design limits for fixed building services and controls in non-domestic buildings

ADL2A contains the following minimum design performance requirements for fixed building services and controls in non-domestic buildings:

- **Energy meters:** To enable at least 90% of each fuel to be assigned to heating, lighting, etc, in accordance with *Building energy metering*[24]; separate meters for low or zero carbon sources; automatic meter recording for buildings with a floor area over 1000 m^2.

- **Heating and hot water system, air handling plant, cooling plant and controls:** Compliance with the *Non-domestic heating, cooling and ventilation compliance guide* [25]; specific fan power at 25% design flow rate no greater than that achieved at 100% design flow rate (that is, in variable volume systems the efficiency of fans at 25% flow rate to be no worse than at 100%); variable speed drives for ventilation system fans (but not smoke control fans) over 1.1 kW; ventilation ductwork airtight in accordance with *Specification for sheet metal ductwork*[26].

- **Controls for heating, ventilation and air conditioning:** Separate control zones for areas of building with significantly different solar exposure, or pattern or type of use.

- **Insulation of pipes, ducts and vessels:** Compliance with the *Non-domestic heating, cooling and ventilation compliance guide*[25].

- **General lighting efficacy in office, industrial and storage areas:** Lighting over all such areas to have average initial efficacy of not less than 45 luminaire-lumens/circuit-watt. (Note that this includes a correction for the light output ratio of the luminaire.)

- **General lighting efficacy in all other types of space:** Average initial efficacy to be not less than 50 lamp-lumens/circuit-watt.

---

[24] *Building energy metering*, TM39, CIBSE, 2006
[25] *Non-domestic heating, cooling and ventilation compliance guide*, ODPM, 2006
[26] *Specification for sheet metal ductwork*, DW144, HVCA, 1998

(Note that this does not include a correction for the luminaire so is a less stringent recommendation.)

- **Lighting controls for general lighting in all types of space:** Local manual switches to be provided, and optionally automatic controls to switch off lighting if there is sufficient daylight or there are no occupants, or following the guidance in *Selecting lighting controls*[27]. The calculation of the BER takes account of reductions in power consumption achieved with automatic lighting controls.

- **Display lighting and controls:** Display lighting to have average initial efficacy of not less than 15 lamp-lumens/ circuit-watt; timers to switch off display lighting when not required.

Criterion 3 – Building to have passive control measures to limit solar gain

ADL1A and ADL2A contain guidance on how to limit overheating in buildings caused by solar gain. This can be done by an appropriate combination of:

> ADL1A and ADL2A contain guidance on how to limit overheating in buildings caused by solar gain

- window size and orientation
- solar protection through shading and other solar control measures
- ventilation (day and night), and
- high thermal capacity.

The Approved Documents refer to the following for further guidance:
- *Reducing overheating – a designer's guide*[28]
- *Climate change and the indoor environment*[29]
- *Solar shading of buildings*[30]
- *Natural ventilation in non-domestic buildings*[31]
- *Use of rooflights to satisfy the 2002 Building Regulations for the Conservation of Fuel and Power*[32]

---

[27] *Selecting lighting controls*, BRE Digest 498, 2006

[28] *Reducing overheating – a designer's guide*, CE129, EST, 2005

[29] *Climate change and the indoor environment*, TM36, CIBSE, 2005

[30] *Solar shading of buildings*, BR 364, BRE, 1999

[31] *Natural ventilation in non-domestic buildings*, AM10, CIBSE, 2005

[32] *Use of rooflights to satisfy the 2002 Building Regulations for the Conservation of Fuel and Power*, NARM, 2002

- *Design for improved solar shading control*[33].

Dwellings. SAP 2005 does not calculate energy use for cooling in dwellings, but does check solar gain to minimise the need for cooling. Dwellings should be designed so that a SAP 2005 assessment indicates that they will not have a high risk of overheating.

Non-domestic buildings without comfort cooling. For non-domestic buildings, ADL2A contains the following performance requirements for limiting solar gain in spaces not served by comfort cooling systems:
- the combined solar and internal casual gains averaged over the period of daily occupancy shall be not greater than 35 W/m^2 when subject to the solar irradiances in CIBSE Guide A, *Environmental design*[34] for the period 0730 to 1730 BST in the month of July; or
- the operative dry resultant temperature shall not exceed 28°C for more than a reasonable number of occupied hours per year. What is reasonable will depend on the activities within the space.

CIBSE TM37 *Design for improved solar shading control* (DISSCO)[33] provides a method for calculating solar gains.

Non-domestic buildings with comfort cooling. Reasonable provision for controlling solar gain in non-domestic building spaces served by comfort cooling is shown by meeting the TER. If solar and internal gains are limited to 35 W/m^2 as above, cooling requirements will be moderate and it will easier to achieve the TER. The notional building assumes modest amounts of glazing (only the area of display windows is the same as in the actual building), so buildings that allow greater solar gain will have to compensate through better energy efficiency in other aspects of the design.

Daylighting. The Approved Documents do not specify minimum daylighting requirements, but the amount of daylight will have an impact on solar gains, conduction heat losses and the need for electric lighting and so will affect predicted carbon dioxide emission rates (the DER and BER). The *Code of practice for*

---

[33] *Design for improved solar shading control*, TM37, CIBSE, 2006
[34] *Environmental design*, Guide A, CIBSE, 1999

daylighting[35] gives guidance on maintaining adequate levels of daylighting.

Quality of construction and commissioning

Criterion 4 — As-built performance to be consistent with design

Regulation 7 of the Building Regulations requires building work to be carried out with adequate and proper materials and in a workmanlike manner. In effect it calls for the finished building to be consistent with the approved design.

> Regulation 7 requires building work to be carried out with adequate and proper materials and in a workmanlike manner

To check that the finished building is consistent with the design, ADL1A and ADL2A require the following:
- a report produced by SAP 2005 or SBEM, as part of the initial calculation of the DER/BER, highlighting the critical design features for use by building control bodies carrying out inspections
- a second calculation of the DER/BER to reflect any changes in performance between design and construction, and incorporating actual measurements (where relevant) of air permeability, ductwork leakage and commissioned fan performance.

In addition, ADL1A and ADL2A require:
- the building fabric to be reasonably continuous insulation (without thermal bridging) over the whole building envelope, as demonstrated by the use of 'approved design details' in combination with on-site inspection
- the building fabric to be reasonably airtight, as demonstrated by pressure testing (where required)
- the building services to be properly installed, as demonstrated by inspection and commissioning reports.

Continuity of insulation
The building fabric should be constructed so that there are no reasonably avoidable thermal bridges in insulation layers – caused by gaps within the various elements, at the joints between elements and at the edges of elements such as those around window and door openings.

---

[35] BS 8206-2. *Lighting for buildings. Code of practice for daylighting*

CONSTRUCTION OF NEW BUILDINGS

ADL1A and ADL2A say reasonable provision would be to:
- adopt approved design details, such as those set out in *Limiting thermal bridging and air leakage: robust construction details for dwellings and similar buildings* [36]
- **or** demonstrate that the details specified by the builder deliver an equivalent level of performance, using the guidance in BRE IP1/06, *Assessing the effects of thermal bridging at junctions and around openings* [37]
- **and**, in addition, demonstrate that an appropriate system of on-site inspection is in place to achieve the required standards of consistency. BRE IP1/06 contains checklists for use in reports (preferably signed by a suitably qualified person) showing the results of on-site inspections of approved details.

Air permeability and pressure testing

Air permeability is the air leakage rate per unit area of the building envelope at a reference internal-to-external pressure difference of 50 Pa, measured in units of m^3/hm^2.

The maximum design air permeability allowed by Part L when doing the first calculation of the DER or BER for the proposed building is 10 m^3/hm^2 at 50 Pa. This is the value assumed in the calculation of the TER based on the notional building. A design air permeability better than 10 m^3/hm^2 will make it easier to meet the TER.

Except in defined circumstances (see below), Regulation 20B requires a completed building to have its air permeability measured by pressure testing, and a pressure test certificate given to the local authority. The measured value of air permeability is then used in the final calculation of the DER or BER in place of the design value.

A satisfactory result would be an air permeability not greater than 10 m^3/hm^2 at 50 Pa, **and** a final DER or BER (using the measured value of air permeability) that is not greater than the TER. These are the only two conditions for a satisfactory test – the measured air permeability does not necessarily need to match the design air permeability.

---

[36] *Limiting thermal bridging and air leakage: robust construction details for dwellings and similar buildings*, ODPM, 2006

[37] *Assessing the effects of thermal bridging at junctions and around openings*, IP1/06, BRE, 2006

The approved procedure for pressure testing is given in *Measuring air permeability of building envelopes*[38]. Pressure testing should be carried out by a suitably qualified person – for example a member of the Air Tightness Testing and Measurement Association (ATTMA).

The pressure test certificate submitted to the local authority should record both the design and the measured air permeability.

Sample pressure testing of dwellings. Not all dwellings that form part of a large development (including a block of flats) need to be pressure tested. But for each development, an air pressure test should be carried out:
- on one unit of each dwelling type if approved construction details have been adopted
- on two units (or 5% if greater[39]) of each dwelling type if approved construction details have **not** been adopted.

To enable lessons to be learned, about half of the scheduled tests for each dwelling type should be carried out during construction of the first 25% of each dwelling type.

When pressure testing is not required. Pressure testing of completed buildings is not mandatory for:
- small developments with no more than two dwellings – but only if the developer can demonstrate that during the preceding 12-month period a dwelling of the same type constructed by the same builder has passed a pressure test
- small developments with no more than two dwellings, or other buildings with a floor area below 500 m² – but only if a high design air permeability of 15 m³/hm² at 50 Pa is assumed when calculating the DER or BER. This will require the building fabric and services to have a better energy performance than if a lower value of air permeability were achieved and demonstrated by pressure testing.

ADL2A contains further guidance relating to the air permeability of the following building types:
- factory-made modular buildings

---

[38] *Measuring air permeability of building envelopes*, TS1, ATTMA, 2006

[39] If the first five dwellings tested all meet the air tightness standard, the figure becomes 2%

CONSTRUCTION OF NEW BUILDINGS

- large extensions to non-domestic buildings whose compliance is being assessed as if they were new buildings – see later) where sealing off the extension from the existing building is impractical
- large complex buildings and compartmentalised buildings.

Consequences of failing a pressure test. If satisfactory performance is not achieved, remedial measures should be carried out on the building and the pressure test repeated until the result is satisfactory.

> If satisfactory performance is not achieved, remedial measures should be carried out and the pressure test repeated

To give builders time to develop the necessary expertise, there are transitional arrangements that apply up to 31 October 2007. For dwellings, and other buildings under 1000 m^2, if a satisfactory air permeability is not achieved – that is, if the measured air permeability is greater than 10 m^3/hm^2 at 50 Pa, or if the DER or BER using the measured value of air permeability is greater than the TER – then the required air permeability is relaxed, but in such a way that reasonable improvements in air leakage will still need to be made.

During the transition period following an unsatisfactory pressure test:
- the new air permeability target for the pressure test becomes:
  - the design air permeability plus 25% of the difference between the initial test result and the design air permeability
  - or, if less demanding, the design air permeability plus 15%
- and the new TER is obtained by substituting the new air permeability target for the reference value of 10 m^3/hm^2 at 50 Pa used originally. The ADs give details of the procedure to follow.

Air leakage testing of ductwork

Air leakage testing of ductwork in non-domestic buildings should be carried out in accordance with the procedures set out in *A practical guide to ductwork leakage testing*[40] on systems served by fans with a design flow rate greater than 1 m^3/s.

---

[40] *A practical guide to ductwork leakage testing*, DW143, HVCA, 2000

Tests should be carried out on the sections of ductwork recommended in the guide, but only if the BER calculation assumes a rate of leakage that is lower than the standard for its pressure class in *Specification for sheet metal ductwork*[41].

If a ductwork system fails to meet the leakage standard, remedial work should be carried out as necessary and new sections tested.

Testing should be carried out by persons with suitable qualifications – for example a member of the HVCA Specialist Ductwork Group or of the Association of Ductwork Contractors and Allied Services.

Commissioning of building services systems

Regulation 20C states that all fixed building services should be properly commissioned and a certificate issued to the local authority.

> All fixed building services should be properly commissioned and a certificate issued to the local authority

Dwellings. In the case of dwellings, the approved commissioning procedure for heating and hot water systems and their controls is given in the *Domestic heating compliance guide*[42].

The certificate issued to the local authority should be signed by a suitably qualified person – for example a person registered with an authorised competent person scheme.

Non-domestic buildings. In the case of non-domestic buildings, the approved commissioning procedure covering the overall process for all building services is given in *Commissioning management*[43].

The certificate issued to the local authority should confirm that:
- a commissioning plan has been followed so that every system has been inspected and commissioned in an appropriate sequence and to a reasonable standard
- the results of tests confirm that the performance is reasonably in accordance with the proposed building designs, justifying any proposals to accept excursions

---

[41] *Specification for sheet metal ductwork*, DW144, HVCA, 1998

[42] *Domestic heating compliance guide*, ODPM, 2006

[43] *Commissioning management*, Commissioning Code M, CIBSE, 2003

The certificate should be signed by a suitably qualified person – for example a member of the Commissioning Specialists Association or Commissioning Group of the HVCA in the case of HVAC services, or of the Lighting Industry Commissioning Scheme in the case of lighting.

Providing information

Criterion 5 – Information to be provided for energy efficient operation

Part L Requirement L1(c) calls for the owner of a building to be provided with sufficient information about the building, the fixed building services and their maintenance requirements to enable the building to be operated efficiently.

Dwellings. For dwellings, operating and maintenance instructions should be provided that are suitable for inclusion in the Home Information Pack[44]. The instructions should cover:
- adjustments to the timing and temperature control settings
- required routine maintenance.

A SAP rating must also be prepared – as required by Regulation 16 even before the 2006 changes to Part L, and fixed in a conspicuous place in the dwelling before it is occupied.

Non-domestic buildings. For non-domestic buildings, the owner should be provided with details of the installed building services plant and controls, their methods of operation and maintenance, and other details required for efficient operation.

> The owner should be provided with details of the installed building services plant and controls, their methods of operation and maintenance, and other details

Suitable guidance is contained in the *Building log book toolkit*[45], which contains standard templates for presenting information.

The information used in the calculation of the TER and BER should be included in the log book.

---

[44] See Housing section at *www.odpm.gov.uk*
[45] *Building log book toolkit*, TM31, CIBSE, 2003

Model designs

Some builders may prefer to adopt model design packages rather than to engage in design for themselves. These model packages of fabric U-values, boiler seasonal efficiencies, window opening allowances, etc, will have been shown to meet the requirements of Part L. Model designs will be made available at *www.modeldesigns.info*.

8 Work in existing buildings

This Chapter presents an overview of the guidance in Approved Documents ADL1B and ADL2B covering work in existing dwellings and existing non-domestic buildings.

Again, this Chapter is not intended to replicate the guidance in the Approved Documents, but to highlight the key requirements. It provides further explanation where desirable, and explains the differences between the requirements for dwellings and other buildings.

The Approved Documents should be consulted for the full details.

The Chapter begins with a description of the types of work covered by ADL1B and ADL2B. It then gives the requirements for work on controlled fittings and services and on thermal elements; and the specific requirements for extensions, changes in energy status, material changes of use, material alterations and consequential improvements.

General guidance

Types of work covered

The 'Definitions' section in Chapter 6 explains when buildings should be treated as dwellings and when as non-domestic buildings.

Generally the requirements are elemental, although for additional flexibility there is an option to use SAP 2005, SBEM or other approved calculation tool to show compliance with Part L for extensions and material changes of use. Also large extensions to non-domestic buildings are to be treated as new build.

ADL1B and ADL2B give guidance on carrying out the following types of work in existing buildings:
- provision of a controlled fitting
- provision or extension of a controlled service
- provision or renovation of a thermal element
- extensions
- changes in energy status and material changes of use
- material alterations
- consequential improvements.

The term 'provision' here means the installation of a new or a replacement item.

'Consequential improvements' are cost-effective improvements to parts of existing buildings that are not affected by proposed building work. The Regulations call for consequential improvements to be made to existing non-domestic buildings only if their floor area is greater than 1000 m^2.

Notifying building work to the local authority. Notification[46] means:
- submitting 'full plans' to the local authority for approval before work starts – the appropriate route for large projects, **or**
- giving a 'building notice' to the local authority at least two working days before starting work – which may be more appropriate for smaller projects, **or**
- presenting an 'initial notice' to the local authority through a private approved inspector before starting work.

Proposed building work does not need to be notified to the local authority when:
- the work is being carried out by a person registered with a competent person self-certification scheme
- the work involves an emergency repair – for example to a failed boiler or leaking hot water cylinder
- the building work is non-notifiable minor work, as described in Schedule 2B of the Building Regulations.

The Government has made certain types of minor building work non-notifiable where the risks of failure are low and notifi-

---

[46] *Building Regulations Explanatory Booklet*, ODPM, 2006

cation would not be cost-effective or useful. The types of Part L work that do not need to be notified are:
- work on heating, hot water and cooling systems in existing buildings that involves
  - replacing control devices that utilise existing fixed control wiring or pneumatic pipes, such as room thermostats, timers or programmers, cylinder thermostats and motorised control valves
  - replacing distribution system output devices, such as radiators and thermostatic radiator valves in an existing heating circuit
  - replacing a circulating pump
  - providing insulation to pipework
- replacing external doors, where the glazed area is not greater than 50% of their internal face area.
- in existing buildings other than dwellings, providing fixed internal lighting where no more than 100 m² of the floor area of the building is to be served by the lighting

Historic buildings. Protected buildings – that is listed buildings, buildings in conservation areas, and ancient monuments – are exempt from the energy efficiency requirements of the Building Regulations.

Where the local authority's conservation officer so advises, special considerations apply to other historic buildings that are not exempt – such as buildings of local architectural and historical interest, and buildings in national parks, areas of outstanding natural beauty, and world heritage sites.

> Protected buildings are exempt from the energy efficiency requirements of the Building Regulations

The aim when doing work on such buildings should be to improve their energy efficiency to the extent that is practical and reasonable, without prejudicing the character of the building or increasing the risk of long-term deterioration to the building fabric or fittings. Guidance is provided in *Building Regulations and historic buildings*[47].

Large extensions to non-domestic buildings. Where the proposed extension to a non-domestic building has a total floor area that is

---

[47] *Building Regulations and historic buildings*, English Heritage, Interim guidance note, 2002

both over 100 m² and more than 25% of the floor area of the existing building, the work should be treated as the construction of a new building and the guidance in Approved Document ADL2A followed.

Guidance relating to building work

Work on controlled fittings

Table 5 gives minimum, area-weighted U-values (or equivalent window energy rating (WER)[48], which takes account also of solar gain and air leakage around the frame) when carrying out work on controlled fittings in existing dwellings and non-domestic buildings.

The whole unit U-value for replacement windows in existing buildings is the same as in Part L 2002, but for windows in new extensions the requirement is for a lower (better) U-value.

As for new construction:
- display windows in non-domestic buildings are exempt from the requirements
- U-values should be calculated using the methods set out in the 2006 edition of BR 443, *Conventions for U-value calculations*[49]
- in non-domestic buildings with high internal gains, the maximum area-weighted average U-value for windows, doors and rooflights can be increased to 2.7 W/m²K, if it can be shown that this will reduce carbon dioxide emissions.

Work on controlled services

Many of the requirements for work on controlled services in existing buildings are the same as for controlled services in new buildings. Commissioning requirements are also the same. However, in ADL1A and ADL2A, design and commissioning requirements are in separate sections, 'Design standards' and 'Quality of construction and commissioning', while in ADL1B and ADL2B the requirements are brought together in one section, 'Guidance relating to building work'.

---

[48] *Windows for new and existing housing*, CE66, EST, 2006

[49] *Conventions for U-value calculations*, BR 443, BRE, 2006 edition

Table 5 Standards (U-values in W/m²K) for work on controlled fittings in existing buildings

Fitting	U-value or WER of new fitting in extension	U-value or WER of replacement fitting in existing building
Window, roof window and rooflight in dwelling	1.8 whole unit, or 1.2 centre pane	2.0 whole unit, or 1.2 centre pane
Window, roof window and rooflight in non-domestic building	1.8 whole unit, or 1.2 centre pane	2.2 whole unit, or 1.2 centre pane
Window, roof window and rooflight in dwelling and similar residential building	Band D WER	Band E WER
Door in dwelling with > 50% of internal face area glazed	2.2 whole unit, or 1.2 centre pane	2.2 whole unit, or 1.2 centre pane
Pedestrian door in non-domestic building with > 50% of internal face area glazed	2.2	2.2
Other door in dwelling	3.0	3.0
High usage entrance door for people in non-domestic building	6.0	6.0
Vehicle access and similar large door	1.5	1.5
Roof ventilator (including smoke vent)	6.0	6.0

Design and commissioning of fixed building services and controls in existing dwellings

ADL1B contains the following requirements when installing fixed building services and controls in existing dwellings:

- **Heating and hot water system appliances, controls and commissioning:** As for new construction, compliance with the *Domestic heating compliance guide*[50].

---

[50] *Domestic heating compliance guide*, ODPM, 2006

When a primary heating appliance is being replaced with one that uses the same fuel, the efficiency of the new appliance must also not be more than two percentage points worse than the efficiency of the old appliance. If the new appliance uses a different fuel with a different carbon dioxide emission factor, the efficiency of the new appliance is multiplied by *the ratio of the carbon dioxide emission factors for the old and new fuels* when doing the comparison.

The commissioning certificate issued to the local authority should be signed by a suitably qualified person.

- **Mechanical ventilation, including air permeability standards for different ventilation strategies:** As for new construction, compliance with *Energy efficient ventilation in housing. A guide for specifiers*[51].

- **Mechanical cooling:** As for new construction, fixed air conditioners to have an energy classification equal to or better than Class C in Schedule 3 of the labelling scheme adopted under The Energy Information (Household Air Conditioners) (No 2) Regulations 2005[52].

- **Insulation of pipes, ducts and vessels:** As for new construction, compliance with the *Domestic heating compliance guide*[50].

- **Fixed internal lighting:** The requirements for new construction apply when:
  o a dwelling is extended
  o a new dwelling is created from a material change of use
  o an existing lighting system is being replaced as part of re-wiring works.

  The requirements in areas affected by the works are: minimum number of fixed low energy light fittings (lamp luminous efficacy greater than 40 lumens per circuit-watt) to be one per 25 m^2 of floor area, or one per four fixed light fittings whichever is the greater. *Low energy domestic lighting*[53] gives guidance on suitable locations for energy efficient

---

[51] *Energy efficient ventilation in housing. A guide for specifiers*, GPG268, EST, 2006

[52] Statutory Instrument 2005 No 1726

[53] *Low energy domestic lighting*, GIL20, EST, 2006

luminaires. In some cases, it may be more appropriate to install a low energy light fitting in an area that is not part of the building work (for example on a landing when converting a loft into a bedroom).

- **Fixed external lighting:** As for new construction, either low energy light fittings; or lamp capacity not to exceed 150 W, with automatic switch-off when not required (that is, when sufficient daylight, or people not detected).

Design and commissioning of fixed building services and controls in non-domestic buildings

ADL2B contains the following requirements when installing fixed building services and controls in existing non-domestic buildings:

- **Energy meters:** As for new construction, to enable at least 90% of each fuel to be assigned to heating, lighting, etc, in accordance with *Building energy metering*[54], separate meters for low or zero carbon sources; automatic meter recording for buildings with a floor area over 1000 m².

- **Heating and hot water system, air handling plant, cooling plant and controls:** As for new construction, compliance with the *Non-domestic heating, ventilation and cooling compliance guide*[55]; specific fan power at 25% design flow rate no greater than that achieved at 100% design flow rate; variable speed drives for ventilation system fans (but not smoke control fans) over 1.1 kW; ventilation ductwork airtight in accordance with *Specification for sheet metal ductwork*[56].

> When plant is being replaced, the efficiency of the new service should not be less than that of the old

When central boiler, chiller and main air handling plant is being replaced, the efficiency of the new service should not be less than the efficiency of the old service. If the new service uses a different fuel with a different carbon dioxide emission factor, the efficiency of the new service is multiplied by *the*

---

[54] *Building energy metering*, TM39, CIBSE, 2006
[55] *Non-domestic heating, cooling and ventilation compliance guide*, ODPM, 2006
[56] *Specification for sheet metal ductwork*, DW144, HVCA, 1998

ratio of the carbon dioxide emission factors for the old and new fuels when doing the comparison.

When replacing a chiller, practical and cost-effective measures should be taken to reduce cooling loads – for example though improved solar control[57] or more efficient lighting.

- **Controls for heating, ventilation and air conditioning:** As for new construction, separate control zones for areas of building with significantly different solar exposure, or pattern or type of use.

- **Insulation of pipes, ducts and vessels:** Insulation of hot and chilled water pipework and storage vessels, refrigerant pipework and ventilation ductwork in accordance with the guidance in *HVAC compliance guide*[58].

- **General lighting efficacy in office, industrial and storage areas:** As for new construction, lighting over all such areas to have average initial efficacy of not less than 45 luminaire-lumens/circuit-watt.

- **General lighting efficacy in all other types of space:** As for new construction, average initial efficacy to be not less than 50 lamp-lumens/circuit-watt.

- **Lighting controls for general lighting in all types of space:** As for new construction, local manual switches to be provided, and optionally automatic controls to switch off lighting if there is sufficient daylight or there are no occupants, or by following the guidance in *Selecting lighting controls*[59].

To encourage the use of automatic controls that reduce power consumption by switching off luminaires when electric lighting is not required, the light output of luminaires is divided by the 'control factor' shown in Table 6 when calculating the average luminaire-lumens/circuit-watt.

---

[57] *Solar shading of buildings*, BR 364, BRE, 1999

[58] *HVAC compliance guide*, TIMSA, 2006

[59] *Selecting lighting controls*, BRE Digest 498, 2006

Table 6 Luminaire control factors

Control function	Control factor
Luminaire in daylit space — light output controlled by photoelectric switching or dimming control, with or without manual override	0.90
Luminaire in space likely to be unoccupied for significant periods — automatic switch-off when occupants leave, manual switch-on (except where unsafe)	0.90
Both of the above	0.85
None of the above	1.00

- **Display lighting and controls:** As for new construction, display lighting to have average initial efficacy of not less than 15 lamp-lumens/circuit-watt; timers to switch off display lighting when not required.

- **Commissioning of building services:** As for new construction, the approved commissioning procedure covering the overall process for all building services is given in *Commissioning management*[60]. The certificate issued to the local authority should confirm that:
  o a commissioning plan has been followed so that every system has been inspected and commissioned in an appropriate sequence and to a reasonable standard
  o the results of tests confirm that the performance is reasonably in accordance with the proposed building designs, justifying any proposals to accept excursions.

  The certificate should be signed by a suitably qualified person — for example a member of the Commissioning Specialists Association or Commissioning Group of the HVCA in the case of HVAC services, of the Lighting Industry Commissioning Scheme in the case of lighting.

- As for new construction, air leakage testing of ductwork should be carried out — and if necessary remedial action taken — in accordance with the procedures set out in *A*

---

[60] *Commissioning management*, Commissioning Code M, CIBSE, 2003

practical guide to ductwork leakage testing[61] on systems served by fans with a design flow rate greater than 1 m³/s. Testing should be carried out by persons with suitable qualifications – for example a member of the HVCA Specialist Ductwork Group or of the Association of Ductwork Contractors and Allied Services.

Guidance on thermal elements

'Thermal element' means a wall, floor or roof (an opaque element of the building fabric) that separates the internal conditioned space from the external environment, including from a space that is not conditioned such as an unheated garage, storage area or plant room.

Regulation 4 already required before Part L 2006 new thermal elements to comply with Part L. Regulation 4A added work on existing thermal elements involving renovation or replacement (see above).

ADL1B and ADL2B give maximum U-values for thermal elements in existing buildings when they are being:
- installed as new in extensions
- replaced
- renovated, and
- retained.

In all cases, worse (that is, higher) U-values may be acceptable if achieving the standard:
- would reduce the floor area by more than 5%
- is technically or functionally not feasible – for example, adding insulation to floors would create difficulties with adjoining floor levels, or the weight of additional insulation might not be supported by the existing structural frame.

New and replacement thermal elements

Maximum U-values for a new thermal element in an extension and for a replacement thermal element in an existing building are given in Table 7.

For new and replacement thermal elements, the U-value of any part of the element should not exceed the limiting value given in Table 10 (see later).

---

[61] *A practical guide to ductwork leakage testing*, DW143, HVCA, 2000

A U-value greater than 0.35 when replacing walls may be acceptable if meeting the standard would result in a reduction in the floor area bounded by the wall of more than 5%.

As for new buildings, the thermal elements should be constructed using approved design details and building techniques so as to:
- avoid thermal bridges in insulation layers – caused by gaps within the various elements, at the joints between elements, and at the edges of elements such as those around window and door openings
- reduce unwanted air leakage through the new envelope parts.

A report should be submitted by a suitably qualified person confirming that:
- either appropriate construction details have been adopted, as specified for example in:
  - *Limiting thermal bridging and air leakage: robust construction details for dwellings and similar buildings*[62]
  - *Guidance for design of metal roofing and cladding to comply with Approved Document L2*[63]
- **or** that the details specified by the builder deliver an equivalent level of performance, using the guidance in BRE IP1/06, *Assessing the effects of thermal bridging at junctions and around openings*[64]
- **and**, in addition, an appropriate system of on-site inspection is in place to achieve the required standards of consistency and conformity with the specifications. BRE IP1/06 contains checklists for use in reports showing the results of on-site inspections of accredited details.

Renovation of thermal elements

Where more than 25% of the surface area of a thermal element is being renovated, the thermal performance of the whole element should be improved to achieve:
- either the standard for replacement thermal elements in Table 7

---

[62] *Limiting thermal bridging and air leakage: robust construction details for dwellings and similar buildings*, ODPM, 2006

[63] *Guidance for design of metal roofing and cladding to comply with Approved Document L2*, MCRMA Technical Paper No 17, 2006

[64] *Assessing the effects of thermal bridging at junctions and around openings in the external elements of buildings*, IP1/06, BRE, 2006

- or, as a minimum, the standard that is practicable and cost-effective within a simple payback of 15 years.

Appendix A of ADL1B gives numerous examples of cost-effective U-value targets when renovating thermal elements.

Table 7 Standards (U-values in W/m²K) for new and replacement thermal elements in existing buildings

Thermal element	U-value of new thermal element in extension	U-value of replacement thermal element in existing building
Wall	0.30	0.35
Floor	0.22	0.25
Pitched roof — insulation at ceiling level	0.16	0.16
Pitched roof — insulation between rafters	0.20	0.20
Flat roof or roof with integral insulation	0.20	0.25

Retained thermal elements

Part L applies to an existing retained thermal element that:
- is part of a building subject to a material change of use (see later)
- is to become part of the thermal envelope and is to be upgraded
- is being upgraded as a consequential improvement.

Retained thermal elements should be upgraded if their U-value is worse than the threshold given in Table 8 to achieve either:
- the improved U-value given in Table 8, or
- as a minimum, the standard that is practicable and cost-effective within a simple payback of 15 years.

Extension of a building

The extension of a building counts as work in an existing building, except in the case of a large extension to a non-domestic building which should be treated as a new building (see below).

Table 8 Standards (U-values in W/m²K) when upgrading retained thermal elements

Thermal element	Threshold U-value	Improved U-value
Cavity wall in dwelling and non-domestic building	0.70	0.55
Other wall type	0.70	0.35
Floor in dwelling	0.70	0.25
Floor in non-domestic building	0.35	0.25
Pitched roof — insulation at ceiling level	0.35	0.16
Pitched roof — insulation between rafters	0.35	0.20
Flat roof or roof with integral insulation	0.35	0.25

Fabric and building services standards

The performance standards described earlier for work in existing dwellings and other buildings apply to work in extensions on:
- controlled fittings
- controlled services
- newly constructed thermal elements
- existing opaque fabric elements that become part of the thermal envelope.

Area of openings in a dwelling. The area of windows, roof windows and doors in dwelling extensions should in most circumstances not exceed the sum of:
- 25% of the floor area of the extension and
- the area of any windows or doors which no longer exist or are no longer exposed with the extension.

Where different approaches need to be adopted to achieve a satisfactory level of daylighting, *Code of practice for daylighting*[65] gives guidance.

Area of openings in a non-domestic building. The area of windows and rooflights in non-domestic building extensions should not exceed:

---

[65] BS 8206-2. *Lighting for buildings. Code of practice for daylighting*

- the proportions given in Table 9
- or, if greater, the proportion of glazing in the part of the building to which the extension is attached.

Table 9 Limits for areas of openings in extensions

Building type	Area of windows and personnel doors as % of exposed wall	Area of rooflights as % of roof
Dwelling	Area of old openings + 25% of floor area of extension	
Residential building	30	20
Place of assembly, office or shop	40	20
Vehicle access doors and display windows and similar glazing	As required	N/A
Smoke vents	N/A	As required

Conservatories and substantially glazed extensions

For a substantially glazed extension to be treated as a conservatory, at least three-quarters of the roof and half of the external walls must be translucent. There must also be effective thermal separation from the heated area in the existing building, with the walls, windows and doors between the building and the extension insulated and draught-stripped to at least the same extent as in the existing building.

Conservatories at ground level and with a floor area no greater than 30 m² are exempt from Part L

Conservatories built at ground level and with a floor area no greater than 30 m² are exempt from Part L of the Building Regulations.

Otherwise, the requirements are that:
- any heating system should have independent temperature and on-off controls, and any heating appliance should comply with the standards for work on controlled services set out earlier
- glazed elements should comply with the standards for replacement fittings in Table 5

- opaque elements should comply with the standards for replacement thermal elements in Table 7.

A proposed extension with substantial glazing may not have sufficient translucent material to qualify as a conservatory, but may otherwise meet the requirements for a conservatory. In this case, a way of complying with Part L would be to show that the area weighted U-value of the elements in the proposed extension is no greater than if it were a conservatory.

Large extensions to non-domestic buildings

If the total floor area of an extension to a non-domestic building is both:
- greater than 100 m², and
- greater than 25% of the floor area of the existing building.

then it should be treated as a new building and follow the guidance in ADL2A.

Optional approaches with more design flexibility

There are two alternative approaches to designing extensions that offer more flexibility than simply to comply with the minimum performance standards for fittings, thermal elements and opening areas given above. They both involve comparing the proposed extension with a notional extension of the same size and shape that complies with the U-values in Table 5 and the opening areas in Table 9.
- First, the U-values given in Table 5 and the opening areas in Table 9 may be varied in the proposed extension provided that the area weighted U-value of all the elements in the extension is no greater than that of the notional extension
- Second, even greater flexibility can be achieved by showing – using SAP 2005, SBEM or another accredited calculation tool – that the carbon dioxide emission rate for the whole of the building and its proposed extension is no greater than that for the building with the notional extension. Any upgrades to the existing building must comply with the relevant guidance in the ADs.

As for new construction, if either of these approaches is adopted then:

- the U-value of any individual element must be no worse than the limiting value in Table 10 (to minimise condensation risk)
- the area weighted average U-value of all the elements must be no worse than the standard in Table 10.

Table 10 Limiting standards (U-values in W/m²K) in extensions

Element	Area-weighted average U-value	Individual element U-value
Wall	0.35	0.70
Floor	0.25	0.70
Roof	0.25	0.35
Windows, roof windows, rooflights and doors	2.2	3.30

Change of energy status and material change of use

Regulation 4B requires buildings subject to a change of energy status to comply with the applicable requirements of Part L of Schedule 1; and Regulation 6 requires buildings subject to a material change of use to comply with the requirements of various Parts of Schedule 1 including Part L.

This means that when changing the energy status of a building or carrying out a material change of use, the above Part L rules apply to:
- the provision or extension of controlled services and fittings
- the provision, renovation and retention of thermal elements

Any existing window, roof window or rooflight that separates a conditioned space from an unconditioned space or the external environment should be replaced according to the above rules if its U-value is worse than 3.3 W/m²K.

To provide more design flexibility, SAP 2005 or other accredited energy calculation method can be used to show that the carbon dioxide emissions from the building as it will become are no greater than if the building had been improved according to the above rules.

Material alteration

Material alterations are defined in Regulation 3. When carrying out a material alteration, the above rules apply to:
- the provision or extension of controlled services and fittings
- the provision, renovation and retention of thermal elements.

Consequential improvements

Regulation 17D requires additional, cost-effective consequential improvements to be made to an existing building with a total useful floor area over 1000 m² when:
- building an extension
- installing fixed building services for the first time
- extending the installed capacity of an existing fixed building service.

The consequential improvements only need to be carried out if they will achieve a simple payback period normally not greater than 15 years, and if they are technically and functionally feasible.

Extensions

For extensions, the value of the consequential improvements should not be less than 10% of the value of the proposed works on the extension, as reported by a chartered quantity surveyor or other suitably qualified person. ADL2B gives examples of the types of improvement that would normally be practical and economically feasible:
- upgrading heating, cooling and air-handling systems more than 15 years old by the provision of new plant and controls
- upgrading inefficient lighting systems that serve areas greater than 100 m² by replacing luminaires and improving controls
- installing energy metering following the guidance in *Metering energy use in non-domestic buildings*[66]
- following the rules given earlier for upgrading retained thermal elements (see Table 8)
- following the rules given earlier for replacing existing controlled fittings (windows, roof windows, rooflights or doors – see Table 5) which have a U-value worse than 3.3 W/m²K

---

[66] *Metering energy use in non-domestic buildings*, TM39, CIBSE 2006

- increasing the on-site low and zero carbon energy sources if the existing sources provide less than 10% of on-site energy demand (but here only if the simple payback period is seven years or less).

Building services

When installing a fixed building service for the first time, or increasing the capacity per unit area of an existing building service:
- **first**, the fabric of the building served by the building service should be improved to the extent that is economically feasible. This will help to reduce the size, initial cost and energy consumption of the building service. The cost of doing this does not count as part of the cost of the consequential improvement
- **second**, additional improvements should be made to the building served by the service if they will achieve a simple payback period normally not greater than 15 years and if they are technically and functionally feasible.

ADL2B gives examples of improvements that would normally be economically feasible:

Example 1. If the installed capacity per unit area of a heating or cooling system is increased, following the rules given earlier for upgrading retained thermal elements (see Table 8).

Example 2. If the installed capacity per unit area of a heating system is increased, following the rules given earlier for replacing existing controlled fittings (windows, roof windows, rooflights or doors – see Table 5) which have a U-value worse than 3.3 W/m²/K.

Example 3. If the installed capacity per unit area of a cooling system is increased, and if the area of windows and roof windows (but excluding display windows) exceeds 40% of the facade area, or the area of rooflights exceeds 20% of the roof area, and the design solar load exceeds 25 W/m²:
- improving solar control such that:
  o the design solar load is no greater than 25 W/m², or
  o the design solar load is reduced by at least 20%, or

○   the effective g-value (shading performance)[67] is no worse than 0.3
• for inefficient lighting systems (with an average lamp efficacy of less than 40 lamp-lumens per circuit-watt), reducing the lighting load by upgrading luminaires or controls according to the earlier rules for work on controlled services.

Providing information

Part L Requirement L1(c) calls for the owner of a building to be provided with sufficient information about the building, the fixed building services and their maintenance requirements to enable the building to be operated efficiently. The requirements for providing information when carrying out work in existing buildings are the same as or similar to those for new construction.

> The owner of a building is to be provided with sufficient information about the building, fixed building services and maintenance requirements so it can be operated efficiently

Dwellings. For dwellings, operating and maintenance instructtions should be provided that are suitable for inclusion in the Home Information Pack[68]. The instructions should cover:
• adjustments to the timing and temperature control settings
• required routine maintenance.

Non-domestic buildings. For non-domestic buildings, the building logbook – containing details of the installed building services plant and controls, their methods of operation and maintenance, and other details required for efficient operation – should be brought up to date (or a new one prepared if it does not exist already).

The new or updated logbook should provide details of:
• any new, renovated or upgraded thermal elements
• any new fixed building services
• any new energy meters
• any other details that enable energy consumption to be monitored and controlled.

---

[67] *Design for improved solar shading control*, TM37, CIBSE, 2006
[68] See Housing section at *www.odpm.gov.uk*

Suitable guidance is contained in the *Building log book toolkit*[69], which contains standard templates for presenting information.

---

[69] *Building log book toolkit*, TM31, CIBSE, 2003

Annex 1 Standard Assessment Procedure

Brian Anderson

This Annex describes SAP 2005 – the Government's Standard Assessment Procedure (SAP) for assessing the energy performance of dwellings[70].

The indicators of the energy performance of buildings are:
- energy consumption per unit floor area
- energy cost rating (the SAP rating)
- environmental impact rating (based on carbon dioxide emissions), and
- DER.

SAP rating

The SAP rating is based on the energy costs associated with space heating, water heating, ventilation and lighting, less cost savings from energy generation technologies. It is adjusted for floor area so that it is essentially independent of dwelling size for a given built form. The SAP rating is expressed on a scale of 1 to 100, the higher the number the lower the running costs.

Environmental impact rating

The environmental impact rating is based on the annual carbon dioxide emissions associated with space heating, water heating, ventilation and lighting, less the emissions saved by energy generation technologies. It is adjusted for floor area so that it is essentially independent of dwelling size for a given built form. The environmental impact rating is

---

[70] *www.bre.co.uk/sap2005*

DER

The Dwelling carbon dioxide Emission Rate is a similar indicator to the environmental impact rating, and is used for the purposes of compliance with the Building Regulations. It is equal to the annual carbon dioxide emissions per unit floor area for space heating, water heating, ventilation and lighting, less the emissions saved by energy generation technologies, expressed in $kgCO_2/m^2/year$.

Calculation method

The method of calculating the energy performance and the ratings is set out in the form of a worksheet, accompanied by a series of tables. The method is compliant with the EPBD.

The calculation should be carried out using a computer program that implements the worksheet and is approved for SAP calculations.

(BRE approves SAP software on behalf of the Department for Environment, Food and Rural Affairs (Defra), ODPM, the Scottish Executive, the National Assembly for Wales and the Northern Ireland Department of Finance and Personnel.)

The Standard Assessment Procedure (SAP) is adopted by Government as the UK method for calculating the energy performance of dwellings. The calculation is based on the energy balance taking into account a range of factors that contribute to energy efficiency:
- materials used for construction of the dwelling
- thermal insulation of the building fabric
- ventilation characteristics of the dwelling and ventilation equipment
- efficiency and control of the heating system(s)
- solar gains through openings of the dwelling
- the fuel used to provide space and water heating, ventilation and lighting
- renewable energy technologies.

The calculation is independent of factors related to the individual characteristics of the household occupying the dwelling when the rating is calculated, for example:

- household size and composition
- ownership and efficiency of particular domestic electrical appliances
- individual heating patterns and temperatures.

Ratings are not affected by the geographical location, so that a given dwelling has the same rating wherever it is built in the UK.

The procedure used for the calculation is based on the BRE Domestic Energy Model, which provides a framework for the calculation of energy use in dwellings. The procedure is consistent with the European standards BS EN 832 and BS EN ISO 13790.

SAP was first published by the DOE and BRE in 1993 and in amended form in 1994, and conventions to be used with it were published in 1996 and amended in 1997. A consolidated edition was published as SAP 1998, and a revised version was published in 2001 (SAP 2001).

The present edition is SAP 2005 in which:
- The SAP scale has been revised to 1 to 100, where 100 now represents zero energy cost. It can be above 100 for dwellings that are net exporters
- DER and environmental impact rating together replace the carbon index
- Energy for lighting is included
- Solar water heating has been revised
- Cylinder loss has been revised; manufacturer's data for heat loss becomes the preferred source of cylinder loss
- The effect of thermal bridging is taken into the account
- Additional renewable and energy saving technologies are incorporated
- A method is provided for estimating a tendency to high internal temperature in summer
- Data tables have been updated (e.g. fuel costs, carbon dioxide emissions, boiler efficiency and heating controls, etc)
- The measure of energy is now kWh rather than GJ.

Scope

The procedure is applicable to self-contained dwellings of total floor area (as defined in Section 1 of SAP 2005, Dwelling dimensions) not exceeding 450 m^2.

For flats, it applies to the individual flat and does not include common areas such as access corridors.

Note: Dwellings of floor area greater than 450 m², common areas of blocks of flats such as access corridors, and other buildings (even though used for residential purposes, for example nursing homes) are assessed using procedures for non-domestic buildings (see Annex 2).

Where part of an accommodation unit is used for commercial purposes (for example as an office or shop), this part should be included as part of the dwelling if the commercial part could revert to domestic use on a change of occupancy. That would be applicable where:
- there is direct access between the commercial part and the remainder of the accommodation,
- all is contained within the same thermal envelope, and
- the living accommodation occupies a substantial proportion of the whole accommodation unit.

Where a self-contained dwelling is part of a substantially larger building and the remainder of the building would not be expected to revert to domestic use, the dwelling is assessed by SAP and the remainder by procedures for non-domestic buildings.

Annex 2 Simplified Building Energy Model

Roger Hitchin

Introduction

From April 2006, there has been a fundamental change to the way of dealing with energy requirements for new buildings in Building Regulations in England and Wales (Scotland and Northern Ireland will follow). Instead of having to satisfy particular values for insulation levels, lighting efficiencies and so on, it is necessary to show that the carbon emissions of the building as a whole are below a particular level. This means doing an energy calculation.

Energy calculation tools

For housing, the calculation will be a new version of the Standard Assessment Procedure, SAP 2005, so the procedure will not be too different from the present situation.

For non-domestic buildings (and a few large dwellings), new procedures will be used. The simplest of these is the Simplified Building Energy Model, SBEM. This is intended to be able to deal with most buildings, and can be downloaded from www.ncm.bre.co.uk together with the user interface, iSBEM, and a user guide.

Some commercial software companies will also include SBEM as part of their product package. Other software – for example energy simulation programs – may also be used, provided that it has been accredited as suitable by ODPM.

The calculations take account of the efficiency of heating, lighting and cooling systems as well as the provision of daylighting and the insulation and airtightness of the building.

Target carbon dioxide emissions

The target emissions value for a building is determined by calculating the emissions of a 'notional building' of the same size, shape, use and weather as the proposed building, but with standard values for insulation, heating efficiencies, etc. This is then reduced by a factor defined in the Regulations to set the target. (All the notional building calculations are carried out automatically by the software.)

There will still be minimum values for insulation, boiler efficiency, etc, but in themselves these will not produce a compliant building.

Energy Performance of Buildings Directive

These changes have been introduced so that the UK will comply with the EPBD.

Other measures will be introduced to comply with further requirements of the EPBD: notably 'energy labelling' (strictly speaking 'energy performance certification' of both new buildings and existing ones when they are sold or let). This will provide a means of identifying new buildings that significantly outperform the Building Regulations requirements. It will also enable purchasers or renters to compare the energy efficiency of new and old buildings.

National Calculation Methodology

Strictly, SBEM is the calculation core issued by ODPM (and iSBEM its associated user interface) of a methodology summarised in the diagram on the next page. Some software vendors are embedding SBEM within their own interfaces, and others will use accredited calculation tools – for example detailed simulation software – in place of SBEM.

The activity database (and weather data) have to be used for compliance, whatever calculation core or interface is used. The fabric and service databases are convenient but not mandatory (though the assumptions used for the notional building may not be changed).

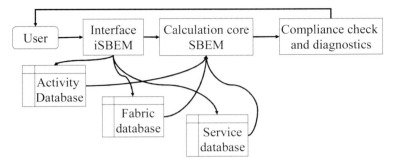

SBEM

In accordance with the requirements of the EPBD, SBEM takes account of:
- indoor conditions such as occupancy, temperature, lighting level, ventilation (set to standard values for Building Regulations purposes)
- building fabric performance (insulation and airtightness)
- energy performance of heating, cooling and hot water services
- lighting and daylighting
- position and orientation
- passive design features
- renewable and CHP options

Wherever possible, its calculation procedures are taken from the emerging CEN standards that support the EPBD. However, in some areas the standards are incomplete, and here new methods built on the principles of the standards have been developed. The resulting software has been independently assessed for ODPM before being released.

Since non-domestic buildings vary enormously in size, usage and design, the software has to be applicable to a wide range of situations. However, to make it suitable for every conceivable building would require more complexity. This would be difficult to achieve without also complicating its use. In particular, the thermal calculation uses a monthly heat balance procedure from EN 13790 (rather than either complex hourly simulations or simple degree-day calculations). This takes account of the impact of the thermal capacity of the structure and can deal with most situations.

In order to check compliance, the user has to describe the shape, orientation and construction of the building, allocate a use to each space and describe the building services systems that are specified. To keep these tasks as straightforward as possible, drop-down lists of options are offered for many decisions – with the user having the freedom to define their own variations if necessary.

Although the immediate application of SBEM is to compliance of new buildings with Building Regulations, it has been developed with an eye to future application to existing buildings as part of the EPBD requirement for Energy Performance Ratings when buildings are sold or rented out. To simplify this potentially time-consuming task, it contains a number of inference procedures that can speed the data entry process, provided that some uncertainty is acceptable.

The current version of SBEM is capable of development to address a number of building and system features that are not presently dealt with, so upgrades can be expected from time to time.

Annex 3 Avoiding overheating

Paul Littlefair

Dwellings

ADL1A 2006 for dwellings contains a requirement to limit 'heat gains'. The requirement applies to all new dwellings, even those where full air conditioning or comfort cooling is already planned. It does not apply to extensions or work in existing dwellings.

The AD explains that

> "Provision should be made to limit temperature rise due to solar gains. This can be done by an appropriate combination of window size and orientation, solar protection through shading and other solar control measures[71], ventilation (day and night) and high thermal capacity."

SAP 2005 Appendix P[72] contains a procedure that enables the designer to check on the likelihood of solar overheating. The method takes account of:
- heat gains through windows
- internal gains from lighting, appliances, cooking and people
- gains from hot water storage, distribution and consumption
- heat loss through the fabric
- natural ventilation

---

[71] *Summertime solar performance of windows with shading devices*, BRE Trust Report FB9, 2005

[72] *The Government's Standard Assessment Procedure for Energy Rating of Dwellings 2005 edition*. Available from *www.bre.co.uk/sap2005*

- average external temperatures (depending on location within the UK)
- thermal capacity of the building.

Further information about techniques to avoid overheating in dwellings can be found in *Reducing overheating – a designer's guide*[73].

Buildings other than dwellings

ADL2A 2006 for non-domestic buildings also contains the requirement to limit "excessive solar gains". It covers new non-domestic buildings, and large extensions (with floor area over 100 m² and more than 25% of the floor area of the existing building).

The Approved Document gives guidance on how to comply with the requirement. It explains that the requirement applies to naturally ventilated spaces as well as those that have mechanical ventilation or cooling. The idea behind this is to avoid the retrofit of cooling systems in naturally ventilated buildings that overheat.

The guidance in the AD only applies to occupied spaces. It mentions three possible design strategies:
- appropriate combination of window size and orientation
- solar protection through shading and other solar control measures, or
- using thermal capacity with night ventilation.

For school buildings, the Approved Document states that Building Bulletins 87[74] and 101[75] specify the overheating criterion and provide guidance on methods to achieve compliance.

For other building types the AD gives two specific ways to comply with the requirement in a space:
- limiting solar and internal casual gain
- showing the space will not overheat.

These are alternatives, so only one of them need be used to demonstrate compliance for a particular space.

---

[73] *Reducing overheating – a designer's guide*, CE129, EST, 2005

[74] *Guidelines for environmental design in schools*, Building Bulletin 87, School Building and Design Unit, DfES, 2003

[75] *Ventilation of school buildings*, Building Bulletin 101, School Building and Design Unit, DfES, 2006

Limiting gains

One way to comply with the requirement is to show that the combined solar and internal casual gain on peak summer days will not be greater than 35 W per m² of floor area in each occupied space. ADL2A explains that the gains are averaged over the time period 0630-1630 GMT (0730-1730 BST). The solar gains are given as the entry for July in the table of design irradiances in Table 2.30 of the 2006 edition of CIBSE Guide A[76]. CIBSE TM37 Designing for improved solar shading control[77] (to be published 2006) explains how to carry out the calculation of gains.

Overheating calculation

Compliance is also possible by showing that the operative temperature in the space does not exceed an agreed threshold for more than a reasonable number of occupied hours per annum. An exact definition of what constitutes overheating is not given in the Approved Document. The threshold temperature, and the maximum number of hours that it is to be exceeded, depend on the activities within the space. The 2006 CIBSE Guide A contains some guidance on this issue.

This is intended to provide a completely flexible method. It could be used, for example, in spaces with night cooling and thermal mass, or where innovative natural ventilation techniques, such as stack effects in tall spaces, are used. The AD does not specify a calculation procedure except to state that the building is tested against the CIBSE Design Summer Year appropriate to the building location. Any reputable calculation technique could be used.

---

[76] *Environmental design*, CIBSE guide A, 2006

[77] *Designing for improved solar shading control*, CIBSE TM37, 2006

Annex 4 Airtightness testing

Mike Jaggs

Introduction

Air leakage is the uncontrolled flow of air through gaps and cracks in the fabric of a building (sometimes referred to as infiltration or draughts). This is not to be confused with ventilation, which is the controlled flow of air into and out of the building through purpose-built ventilators and is required for the comfort and safety of the occupants.

Too much air leakage leads to unnecessary heat loss and discomfort to the occupants from cold draughts.

The increasing need for higher energy efficiency in buildings means that airtightness has become a major performance issue. The aim should be to "Build tight – ventilate right". Taking this approach means that buildings cannot be too airtight, but it is essential to ensure appropriate ventilation rates are achieved through purpose-built ventilation openings.

Assessment of building envelope air leakage involves establishing a pressure differential across the envelope and measuring the airflow required to achieve that differential. This is normally achieved by utilising variable flow portable fans, which are temporarily installed in a doorway, or other suitable external opening.

Airtightness requirements for conserving energy are covered in Part L of the Building Regulations.

Part L air permeability requirements

Part L specifies minimum airtightness performance requirements in terms of air permeability. This is defined as the air leakage in cubic metres per hour per square metre of building envelope area (m^3/hm^2) at a reference pressure of 50 Pa.

The maximum value of air permeability permitted by Part L is 10 m^3/hm^2, but to meet the carbon emissions target specified in Part L it is advisable to aim for an air permeability much lower than this.

A more appropriate design air permeability for meeting the Part L carbon emissions target would be 7 m^3/hm^2, while a best practice design value would be 3 m^3/hm^2.

Part L airtightness testing requirements

The airtightness of all buildings with a gross floor area above 500 m^2 must be tested to demonstrate that their air permeability meets Part L requirements.

For all other buildings, including dwellings, airtightness testing is optional. However, if testing is not carried out a high default air permeability of 15 m^3/hm^2 must be assumed in the carbon dioxide emissions calculation for the building. To compensate for this high air permeability, measures such as thicker insulation or the use of renewable energy systems may be needed to achieve the carbon emissions target.

It is likely to be more cost-effective to design a building for low permeability and to carry out an airtightness test to show that the design air permeability has been achieved.

Guidance

Full guidance on verifying airtightness performance is given in the Airtightness Testing and Measurement Association (ATTMA) document TS1[78], which can be downloaded from *www.attma.org*.

Technical advice and guidance on designing and constructing an airtight building is provided in BRE Report BR448, *Airtightness in commercial and public buildings*[79].

---

[78] *Measuring air permeability of building envelopes*, TS1, ATTMA, 2006

[79] *Airtightness in commercial and public buildings*, BRE Report BR 448, 2002

Annex 5 Low and zero carbon systems

Ken Bromley

The Office of the Deputy Prime Minister (ODPM) has published a guide to low and zero carbon (LZC) energy sources[80] to help builders meet the tighter carbon dioxide emission standards for whole buildings now specified in Part L of the Building Regulations. The guide focuses on microgeneration – defined as the small-scale production of heat and/or electricity from a low carbon source. Part L targets for non-domestic buildings in particular have been set so as to encourage up to 10% of energy supply to come from LZC sources.

This annex highlights the costs and cost-effectiveness of the LZC sources described in the ODPM guide, and their most suitable application.

The ODPM guide reviews eight LZC technologies:

- absorption cooling
- biomass heating
- micro-CHP (combined heat and power)
- ground cooling
- ground source heat pumps (for heating and cooling)
- solar electricity (photovoltaic panels and tiles)
- solar hot water
- wind turbine generators.

The guide shows how to calculate the carbon dioxide savings obtained with an LZC source – essentially the carbon dioxide

---

[80] *Low or zero carbon energy sources: strategic guide*, ODPM, 2006

produced by the conventional energy source less the carbon dioxide (if any) produced by the LZC source.

Micro-CHP units, photovoltaic panels and wind turbines are normally permanently connected to the grid so that any excess electricity they generate can be 'exported'. All the electricity generated by them is therefore assumed to displace grid electricity.

The LZC guide assumes the following 'fuel factors':
- burning natural gas produces 0.194 $kgCO_2/kWh$
- consuming grid electricity produces 0.422 $kgCO_2/kWh$
- displacing grid electricity with a renewable source saves 0.568 $kgCO_2/kWh$

The figure for displacing grid electricity is higher than for consuming grid electricity because it is assumed that renewable energy sources will lead to the least efficient power stations being closed down first.

Although LZC sources save carbon, they are often not cost-effective. Calculations are difficult, but a November 2005 report for the DTI[81] concluded that few microgeneration technologies are cost-effective at present. Notable exceptions were biomass heating and ground source heat pumps, which can be cost-effective compared with electric heating.

A BRE study of the scope for reducing carbon emissions from housing[82] concluded that solar hot water and PV are both currently not cost-effective, although PV could become cost-effective if its costs fall dramatically in coming decades as is being predicted.

Table A1, taken from a recent report[83], shows the current estimated cost of using a range of building-integrated LZC technologies to supply 10% of energy needs to an estate of ten units.

The ODPM guide points out that LZC technologies should be applied to a building only after first taking full advantage of conventional energy efficiency measures.

---

[81] *Potential for microgeneration — study and analysis*, Energy Saving Trust, Econnect, ElementEnergy, November 2005. Available from *www.dti.gov.uk/energy/consultations/pdfs/microgeneration-est-report.pdf*

[82] *The scope for reducing carbon emissions from housing*, BRE IP15/05, 2005

[83] *Renewable energy — a current perspective on planning and design options*, Broadway Malyan and Fulcrum Consulting, July 2005

Table A1 Cost of microgeneration applied to an estate

Technology	Cost/unit (10 blocks)
600 kW wind turbine	£421
Bio-diesel CHP	£803
60 kW wind turbine	£809
Biomass heating	£837
Solar hot water	£1289
Aquifer thermal energy storage (ground source heating and cooling)	£1500
15 kW wind turbines	£1503
6 kW wind turbines	£1874
2.5 kW wind turbines	£2813
Ground source heat pump	£3343
Photovoltaics	£5463

An office complex completed in 2003 and belonging to Renewable Energy Services employs a number of LZC systems. The historical and current performance of the systems can be viewed at the RES website[84].

In March 2006, the Government published its Microgeneration Strategy[85], which aims to create conditions under which microgeneration becomes a realistic alternative or supplementary energy generation source for the householder, community and small businesses. The Energy Saving Trust believes that by 2050 microgeneration could provide 30–40% of the UK's electricity needs and help to reduce household carbon dioxide emissions by 15% per annum[81].

Absorption cooling

Absorption cooling uses heat instead of electricity to provide cooling, and is mainly applied to commercial and industrial buildings. Absorption chillers can be gas fired, or they can use waste heat from a combined heat and power (CHP) plant or from an industrial process.

To illustrate the carbon benefits of using waste heat, the LZC guide shows the amount of CO_2 per kWh of heat input that

---

[84] http://res.esbensen.dk

[85] *Our energy challenge — Power from the people*, DTI, March 2006

might be associated with the following types of absorption cooling:
- gas-fired — 0.226 kgCO$_2$/kWh
- waste heat from a CHP plant generating heat and electricity — 0.039 kgCO$_2$/kWh
- waste heat from a process that would otherwise be rejected — zero kgCO$_2$/kWh

The cooling demand that can be met by the chiller will depend on its coefficient of performance (cooling effect in kW divided by energy input in kW), which can vary from 0.7 for standard units to 1.4 for the most efficient.

Absorption chillers are more expensive than chillers driven by electricity.

Biomass heating

Biomass is a solid fuel consisting most commonly of wood chips, logs or pellets. It is close to being 'carbon neutral', in that the amount of carbon dioxide emitted during the burning process matches the amount of carbon dioxide absorbed by the wood fuel (for example willow) while it was growing.

Biomass has applications ranging from domestic boilers, room heaters and stoves, to large-scale heating boilers and combined heat and power.

Biomass boilers can be very efficient, but currently have a high capital cost, in part due to the need for an automatic fuel feeding mechanism.

A household wood-fuelled boiler system burning logs or pellets to provide 15 kW has a typical cost of £4500[86].

Micro-CHP

Combined heat and power (CHP) plants produce heat and electricity. Large systems are used in hospitals and hotels, while small 'micro-CHP' units – although not yet proven to reduce carbon dioxide emissions – are now becoming available for domestic-scale applications.

CHP can reduce carbon dioxide emissions compared with centrally generated grid-supplied electricity in two ways:

---

[86] *www.clear-skies.org*

- by using the heat from the electricity generating process that would otherwise be wasted, and
- by reducing transmission losses over the electricity grid.

The technologies that micro-CHP units can be based on include:
- internal combustion engines, producing around 5 kW electricity and 10 kW heat
- Stirling (heat) engines, producing around 1 kW electricity and 8 kW heat
- hydrogen fuel cells (still under development).

For energy efficiency, operation must be led by the demand for heat, with a sophisticated control system – micro-CHP units are not energy efficient if used only to generate electricity in the summer, for example.

Micro-CHP units are unlikely currently to be appropriate for modern flats and houses with small heat losses.

Cost estimates[87] are an extra £500 for a Stirling engine micro-CHP unit compared with a conventional gas boiler, and a total installed cost of £12,000 for an internal combustion engine micro-CHP unit in, for instance, a sheltered housing application.

Ground cooling

Ground cooling exploits the relatively low and constant temperature of the ground to provide summertime cooling.

The technology involves either:
- drawing outside air into a building through a network of underground pipes, or
- pumping low temperature underground water through a heat exchanger in the building's ventilation system, or through coils fitted in the floors, ceilings or walls of the building.

Capital costs are relatively high and the technology is most appropriate for commercial buildings.

Ground source heat pumps

Ground source heat pumps are widely used in residential and commercial buildings for space heating, water heating and air conditioning. They use the technology of refrigerators to either

---

[87] *www.cogen.org*

extract heat from the ground and release it at a higher temperature into a building or, working in reverse, to provide cooling. Heat pump compressors are typically driven by electricity.

A system capable of supplying 50% of the space and water heating demand of a typical house would require around 80 to 100 m of ground coils or 'collectors'. Collectors can be arranged horizontally in shallow trenches, or buried in vertical boreholes.

Ground source heat pumps can be economically viable for residential applications where they replace electric heating and hot water and provide 100% of the demand. In non-residential applications, they need to provide at least 50% of the heating and hot water demand to be cost-effective.

The typical cost of a domestic system is £4000 to £6000[88].

Solar electricity

Photovoltaic (PV) cells convert sunlight into direct current electricity. PV panels can be mounted on roofs or walls, or PV tiles are available to replace standard tiles. They are usually connected to the building's alternating current electricity supply through an electronic 'inverter'.

Although PV costs have fallen significantly in recent years, and are likely to continue to fall as the technologies are improved further, solar PV is probably currently one of the least cost-effective of all the renewable technologies.

One square metre of PV costs £500 to £1000 and in the UK can be expected to generate 0.1 to 0.2 kWh per day of electricity. The average household consumption of electricity is around 10 kWh per day.

Solar hot water

In contrast, solar thermal systems are one of the more cost-effective of renewable technologies – particularly where there is a high demand for hot water such as in hospitals, nursing homes and leisure facilities. They can provide up to 50% of domestic hot water needs in both dwellings and other buildings.

They work by using roof- or wall-mounted solar collectors to pre-heat the incoming cold water – by means of a secondary coil fitted in either the main hot water cylinder, or in a second hot water cylinder. Solar thermal panels can be expected to generate

---

[88] *www.clear-skies.org*

around 0.8 kWh of thermal energy per m² per day in UK conditions.

A typical single house system, occupying 2.5 to 4 m² of roof space and costing £1500 to £5000, could supply up to 50% of hot water demand.

Wind turbines

Wind turbines convert wind to electricity[89]. They have rated outputs ranging from 500 W to 1.5 MW, although 500 kW is a practical limit in built-up areas.

The largest turbines are the most efficient and cost-effective. Given reasonably windy conditions of around 13 mph, each installed kW can be expected to produce around 2500 kWh annually at a cost of £2500 to £5000.

In practice, in the 12-month period to March 2005, the RES 225 kW wind turbine located near Watford generated 194 MWh of electricity – or 862 kWh per year per installed kW.

Typical domestic rooftop units with a rated output of 500 W are less efficient than bigger turbines. At average wind speeds of around 13 mph, manufacturers claim an output of over 500 kWh per year. However, power output – as for all wind turbines – depends critically on wind speed, ranging from 40 W at 11–12 mph to 400 W at 37 mph.

At BRE's location near Watford, DTI figures[90] indicate that the average annual wind speed at a height of 10 m is 10.3 mph, and at a height of 45 m it is 13.4 mph, but in a built-up area the wind might be turbulent and affected by obstructions. In practice, therefore, and taking account also of inverter losses, the output from a small roof-top unit in an urban environment might be below 200 kWh per year (0.5 kWh per day).

---

[89] *Small-scale, building integrated wind power systems*, BRE IP 12/05, 2005
[90] *www.dti.gov.uk/renewables/technologies/windspeed*

Annex 6 Domestic heating compliance guide

Introduction

The following text is taken with the permission of the ODPM from the Introduction to the *Domestic heating compliance guide*[91].

The guide is a second tier document referred to in ADL1A and ADL1B as a source of guidance on the means of complying with the requirements of the Building Regulations for space heating and hot water systems. It was prepared with the assistance of industry bodies and covers conventional means of providing primary and secondary space heating and domestic hot water for dwellings in use in England and Wales.

For new dwellings, guidance is provided on the design limits for building services systems referred to in ADL1A. For existing dwellings, guidance is provided on reasonable provision for the installation or replacement of controlled services as referred to in ADL1B.

The guide identifies standards of provision that meet the requirements for systems in new build and in existing buildings when work is being undertaken. The levels of performance for new and existing dwellings differ only where practical constraints arise in existing dwellings. The Domestic Heating Compliance Guide covers a range of frequently occurring situations but alternative means of achieving compliance may be possible. The status of alternative provisions is explained in the 'Use

---

[91] *Domestic heating compliance guide*, ODPM, 2006

of guidance' sections at the front of the Approved Documents.

The guide also refers to third tier publications, which include information on good practice for design and installation over and above the minimum regulatory provision.

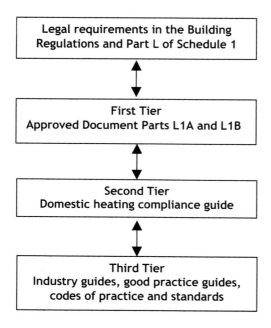

Annex 7: Non-domestic heating, cooling and ventilation compliance guide

Introduction

The following text is taken with the permission of the ODPM from the Introduction to the *Non-domestic heating, cooling and ventilation compliance guide*[92].

The guide is a second-tier document referred to in ADL2A and ADL2B as a source of guidance on the means of complying with the requirements of Building Regulations Part L for space heating systems, hot water systems, cooling and ventilation systems. It was prepared by industry bodies and ODPM and covers the conventional means of providing primary space heating, domestic hot water and comfort cooling and ventilation for buildings in use in England and Wales. When appropriate, the guide identifies the different requirements for systems in new build and those in existing buildings where work is being undertaken.

The guide outlines the minimum provisions for compliance with Part L for each type of heating, hot water, cooling or air distribution system as follows:

---

[92] *Non-domestic heating, cooling and ventilation compliance guide*, ODPM, 2006

- Minimum provisions for efficiency of the plant that generates heat, hot water or cooling
- Minimum provisions for controls to ensure that the system is not generating heat, hot water or cooling unnecessarily or excessively
- A set of additional measures which may improve the efficiency of the plant; these are non-prescriptive and may be either required or optional depending on the type of plant
- Minimum provisions for other factors affecting the safety or energy efficiency of the system
- Minimum provision for insulation of pipes and ducts serving space heating, hot water and cooling systems
- Minimum provisions for acceptable specific fan power ratings for fans serving air distribution systems.

The Building Regulations Part L 2006 requires the energy performance of buildings other than dwellings to be calculated using the NCM[93]. The NCM defines the procedure for calculating the annual energy use for a proposed building (based on a range of factors including the properties of the walls, floors, roofs and glazing as well as the building services) and comparing it with the energy use of a comparable 'notional' building. The NCM also calculates the rate of carbon emissions from the building, which should not be greater than its TER as described in Part L and also calculated by the NCM. The NCM can be implemented through accredited simulation software or through SBEM[94].

The guide identifies the input parameters that are required by the accredited NCM models (for example SBEM) for space heating, hot water, comfort cooling and ventilation systems, in order to calculate the annual energy performance.

The key requirements for compliance with Part L and ADL2A and ADL2B and the key parameters for input into the accredited NCM models (for example SBEM) are summarised in a compliance checklist.

---

[93] *The National Calculation Methodology for Part L*, ODPM, 2006

[94] SBEM tool can be downloaded from *www.odpm.gov.uk*

Abbreviations

ATTMA	Air Tightness Testing & Measurement Association *www.attma.org*
BER	Building Emission Rate
BSI	British Standards Institution *www.bsi-global.com*
CEN	European Committee for Standardization *www.cenorm.be*
CHP	Combined heat and power
CIBSE	Chartered Institution of Building Services Engineers *www.cibse.org*
CWCT	Centre for Window and Cladding Technology *www.bath.ac.uk/cwct*
Defra	Department for Environment, Food and Rural Affairs *www.defra.gov.uk*
DER	Dwelling Emission Rate
DfES	Department for Education and Skills *www.dfes.gov.uk*
EH	English Heritage *www.english-heritage.org.uk*
EPBD	Energy Performance of Buildings Directive
EST	Energy Saving Trust *www.est.org.uk*
HMO	House in multiple occupation
HIP	Home Inspector's Pack
HVAC	Heating, ventilation and air conditioning
HVCA	Heating and Ventilating Contractors Association *www.hvca.org.uk*
iSBEM	Interface tool for the Simplified Building Energy Model
LZC	Low and zero carbon
MPBA	Modular and Portable Building Association *www.mpba.biz*
NARM	National Association of Rooflight Manufacturers *www.narm.org.uk*
NCM	National Calculation Method *www.ncm.bre.co.uk*
NDHC	*Non-domestic heating, cooling and ventilation compliance guide*

MCRMA Metal Cladding and Roofing Manufacturers
        Association *www.mcrma.co.uk*
ODPM   Office of the Deputy Prime Minister *www.odpm.gov.uk*
SAP 2005 Standard Assessment Procedure 2005
SBEM   Simplified Building Energy Model *www.ncm.bre.co.uk*
TER   Target Emission Rate
TRV   Thermostatic radiator valve
TIMSA   Thermal Insulation Manufacturers and Suppliers
        Association *www.timsa.org.uk*
WER   Window energy rating

References

A practical guide to ductwork leakage testing, DW143, HVCA, 2000

Airtightness in commercial and public buildings, BRE Report BR 448, 2002

Assessing the effects of thermal bridging at junctions and around openings, IP1/06, BRE, 2006

BS 8206-2. *Lighting for buildings. Code of practice for daylighting*

Building log book toolkit, TM31, CIBSE, 2003

Building Regulations and historic buildings, English Heritage, Interim guidance note, 2002

Building Regulations Explanatory Booklet, ODPM, 2006. Available from www.odpm.gov.uk

Climate change and the indoor environment, TM36, CIBSE, 2005

Commissioning management, Commissioning Code M, CIBSE, 2003

Conventions for U-value calculations, BR 443, BRE, 2006 edition.

Design for improved solar shading control, TM37, CIBSE, 2006

Designing for improved solar shading control, CIBSE TM37, 2006

Designing with rooflights to satisfy ADL2 (2006), NARM, 2006

Domestic heating compliance guide, ODPM, 2006

Energy efficient ventilation in housing. A guide for specifiers, GPG268, EST, 2006

Energy performance standards for modular and portable buildings, MPBA, 2006

Environmental design, Guide A, CIBSE, 1999

Guidance for design of metal roofing and cladding systems to comply with Part L (2006), MRCMA Technical Paper No 17, 2006

Guidelines for environmental design in schools, Building Bulletin 87, School Building and Design Unit, DfES, 2003

House of Lords Select Committee on Science and Technology Second Report, www.parliament.the-stationery-office.co.uk, July 2005

HVAC compliance guide, TIMSA, 2006

Limiting thermal bridging and air leakage: accredited details for dwellings and similar buildings, ODPM, 2006

Limiting thermal bridging and air leakage: robust construction details for dwellings and similar buildings, ODPM, 2006.

Low energy domestic lighting – a summary guide, EST, GIL20, 2002

Low or zero carbon energy sources: strategic guide. ODPM, 2006

Measuring air permeability of building envelopes, TS1, ATTMA, 2006

Metering energy use in non-domestic buildings, TM39, CIBSE 2006

Natural ventilation in non-domestic buildings, AM10, CIBSE, 2005

Non-domestic heating, cooling and ventilation compliance guide, ODPM, 2006

Our energy challenge – Power from the people, DTI, March 2006

Potential for microgeneration – study and analysis, Energy Saving Trust, Econnect, ElementEnergy, November 2005. Available from *www.dti.gov.uk/energy/consultations/pdfs/microgeneration-est-report.pdf*

Reducing overheating – a designer's guide, CE129, Energy Saving Trust, 2005

Reducing overheating – a designer's guide, CE129, EST, 2005

Renewable energy – a current perspective on planning and design options, Broadway Malyan and Fulcrum Consulting, July 2005

Selecting lighting controls, BRE Digest 498, 2006

Small-scale, building integrated wind power systems, BRE IP 12/05

Solar shading of buildings, BR 364, BRE, 1999

Specification for sheet metal ductwork, DW144, HVCA, 1998

Statutory Instrument 2005 No 1726

Summertime solar performance of windows with shading devices, BRE Trust Report FB9, 2005

The Government's Standard Assessment Procedure for Energy Rating of Dwellings 2005 edition. Available from *www.bre.co.uk/sap2005*

The National Calculation Methodology for determining the energy performance of buildings. Part 1: A guide to the application of the SBEM and other approved calculation tools for Building Regulations purposes, ODPM, 2006

The National Calculation Methodology for Part L, ODPM, 2006.

The scope for reducing carbon emissions from housing, BRE IP15/05, 2005

The thermal assessment of window assemblies, curtain walling and non-traditional building envelopes, CWCT, 2006

Ventilation of school buildings, Building Bulletin 101, School Building and Design Unit, DfES, 2006

Windows for new and existing housing, CE66, EST, 2006